ALUMINUM·COMPOUNDS
in FOOD

ALUMINUM COMPOUNDS *in* FOOD

Including a Digest of the Report of the Referee Board
of Scientific Experts on the Influence of Aluminum
Compounds on the Nutrition and Health of Man

BY

ERNEST ELLSWORTH SMITH, Ph.D., M.D.

Fellow and Former President, New York Academy of Sciences
Fellow of the New York Academy of Medicine, etc., etc.

PAUL B · HOEBER · Inc
NEW YORK · MCMXXVIII

PREFACE

U P TO the present time the results of studies relating to the presence and effects of compounds of aluminum in food have been widely scattered in the scientific literature. The author believes the subject has become of such general scientific interest as to demand its presentation within the pages of a single volume. It is thought that it will be of value not only to those interested in the particular subject but as well to a much larger number interested in scientific methods of investigating the composition and value of food.

This volume records researches by the author which heretofore have not appeared in scientific literature, and gives some details of unpublished work by other investigators, including a comprehensive résumé of the investigations by the Referee Board of the United States Government.

The author makes grateful acknowledgment to those authorities he has freely quoted, whether or not he is in full accord with their methods of research or the conclusions they have reached.

E. E. SMITH.

NEW YORK, N. Y.,
December, 1927.

CONTENTS

and the Conditions of its Possible Absorption. Experiments with
a diet poor in phosphates.
Conclusions.

2. Abstracts of the Testimony of PROF. RUSSELL H. CHITTENDEN. 130

The problems presented to and the participation of the Board.
Bulletin No. 103. Literature. Small and large quantities of aluminum. The work at New Haven; the experiments by Dr. Long and
Dr. Taylor. The results. The cathartic action of other forms of
baking powder. Description of New Haven experiments. Value of
metabolic experiments. Has seen no later results that would lead
to any change in his conclusions, which are a part of the conclusions of the Referee Board. Answers to criticisms as to the use of
healthy subjects; as to the duration of the period of the feeding
experiments; as to possible phosphorus depletion. The question of
the absorption of aluminum.
The work of Thorek and Kahlenberg. Citation from Schmiedeberg's Pharmacology. Value of results obtained by direct immersion of unprotected cells in solutions of aluminum compounds.
The significance of small quantities of aluminum in the blood.
The inhibition of plant growth by aluminum compounds dissolved in acid soils. Citation from Cushny's Pharmacology, as
to the action of alum solutions. Answer to criticism of the form of
aluminum compounds used in Dr. Taylor's experiments.

Norfolk Baking Powder Case. Case, about 1892. Excelsior Baking
Powder Case. State of Missouri vs. Whitney Layton. State of
Missouri vs. The Great Western Coffee and Tea Company. The
People of the State of California vs. (1) J. A. Saint; (2) A. Zimble-
man. Commonwealth of Pennsylvania vs. Meyer Gross. Federal
Trade Commission, Complainant, vs. Royal Baking Powder Co.,
Respondent; Docket #540.

Value of such observations. The work of House and Gies on lupin
seedlings; of Hartwell and Pember on acid soils; of S. D. Conner
on acid soils. The studies of W. J. V. Osterhout on the influence on
conductivity. Studies of Lester C. Himebaugh on the penetration
through a colloidal medium and the antiseptic action, by the agar
plate cup method. Similar studies by E. E. Smith, in criticism of
Himebaugh's and Osterhout's conclusions. The work of Julius
Stoklasa on plant germination and growth. Conclusions.

Earlier studies. Observations of Siem and of Döllken complicated by tartrate administration. Action of tartrates. Aluminum
in feces following intravenous administration (Steel). Subcutaneous
administration to mice (Himebaugh). Intravenous administration
to rabbits (Seibert): changes in blood morphology; increased resistance to hemolyzing agents; comparison with administration by
mouth; effects on anti-body production.

The claim that alumed flour was injurious. The meaning of unwholesomeness or injury to health. The value and limitations of
studies under unnatural conditions and with abuse in experimental
administration. Experimental studies by Leary and Sheib; by
Gies and coworkers on the albino rat, as to growth, production of
symptoms of toxicity, fecundity, health of offspring; similar studies
by Hawk and co-workers. The author's analysis of data relating
to the foregoing diets. Experiments by E. E. Smith. Comparisons
of results. Criticism of the experiments of Gies, of Hawk and
of Smith by Steenbock; by Mendel. Research of Daniels and
Hutton indicating a biogenic function of aluminum compounds;
confirmatory observations by Mitchell and Schmidt.

Introduction. Claims of unwholesomeness: by deflecting phosphates;
by depriving the body of phosphates; by diminishing urinary
acidity; by combining with food accessory substances; by acting
as irritants in the gastrointestinal tract; by impairing digestive
efficiency. The effects of aluminum compounds administered independent of food; of massive doses of baking powder residue in
food. Inhibition of enzyme activity. Alleged production of a gastrointestinal putrefactive toxemia. The claim that aluminum is
absorbed and accumulates in the body; the possibility that it is a
normal constituent of the body. Comparison with iron. Tests for:
without prior feeding with aluminum compounds; with prior
administration of soluble compounds of aluminum; with prior
administration of insoluble compounds of aluminum; with prior
feeding of food prepared with S. A. S. baking powder. The R. B. units
of aluminum ingestion. Aluminum as a protoplasm poison. The claim
that absorbed aluminum has a deleterious effect. Protective factors.
The claims of the production of toxic symptoms; of retarded

ALUMINUM COMPOUNDS IN FOOD

I

INTRODUCTION

COMPOUNDS OF ALUMINUM may occur in food either as natural or as added ingredients. That they are commonly present as natural ingredients is not surprising when we consider the abundance of aluminum in the earth's crust. The average elemental composition of known terrestrial matter, including the solid earth, the waters of the earth and the surrounding atmosphere, has been estimated[1] to be: oxygen 50.02 per cent, silicon 25.80 per cent, aluminum 7.3 per cent and iron 4.18 per cent. Hence, aluminum ranks third in abundance and is more plentiful than any other metal, comprising as it does nearly 8 per cent of the earth's crust, nearly twice the content of iron or calcium and three times that of sodium or magnesium.

These facts are of importance since they show that in the adjustment of living matter to its environment there has necessarily been intimate contact with compounds of aluminum. Hence, it would be expected to follow either that aluminum is not inimical to the chemical and physical processes that constitute life, or that living matter has come to be protected from such intimate contact as would be inimical, even from the aluminum that occurs so widely as a natural ingredient of foods. From the following pages, it will be seen that there is evidence that in relatively very small amounts aluminum compounds exercise a biogenic function, certainly in some living structures. As regards compounds of aluminum naturally occurring in foods, there is evidence that the higher forms of life are so constituted as to protect their tissues from any degree of contact that might prove injurious.

In the consideration of aluminum compounds as added ingredients in foods the essential problem is to determine the effects of such foods upon the health of man. Before dealing with that problem, however, it should be determined what, if any, purpose is served by these added aluminum compounds, for if such additions are merely for sophistication, their presence cannot be approved, irrespective of whether or not the foods are wholesome.

To cucumber pickles, to maraschino cherries and to some accessory food materials, alum, aluminum sulphate or some other aluminum compound is added in very small amount for the purpose of rendering the preserved article crisp and firm like the fresh product, hence, more appetizing. The amount of aluminum ingested as a result of such addition is extremely small and, considering the varying amount consumed as naturally occurring ingredients, ordinarily does not materially increase the total ingested.

Aluminum compounds are also added to foods as the acid ingredient of certain classes of baking powders. Since these latter serve the useful purpose of leavening foods, an unquestionably important process in its preparation, the matter of the wholesomeness to man of food prepared with such baking powders is of economic as well as of hygienic importance. As will be seen, the amount of compounds of aluminum added to food from baking powders will ordinarily slightly increase aluminum ingestion; however, this is only a fraction of what can be ingested in the form of aluminum compounds occurring naturally.

Since the addition of aluminum compounds as ordinarily practiced would seem to serve a useful purpose in the manufacture of foods emphasis will be given here to the question of wholesomeness. The fact that a particular ingredient, be it aluminum residue from baking powder or any other constituent of the food, is capable of producing injurious effects, either by introduction through unnatural channels or by what the writer has elsewhere[2] termed abuse of the quantitative relation, does not solve the problem. Practically any food or food ingredient may be made unwholesome by abuse. Therefore, having determined the modes of action experimentally by abuses of administration or quantity consumption the problem is to discover whether like actions

would result from normal ingestion as an added food ingredient. If they would, the ingredient renders the food unwholesome; if they would not, and if the quantity is removed by a
wide margin of safety from one that might so act, then it
does not render the food unwholesome.

In the following chapters are presented data on the
natural occurrence of aluminum compounds in, and their
addition to, food, on the effects upon the body of their
abuse, both as to channels of administration and as to form
and amount ingested, and on their effects when ingested as
normal ingredients of food, following which, in the last
chapter, is a discussion of the significance of these data and
their interpretation.

In the presentation of data and conclusions of others,
their statements are freely presented, frequently using substantially their own language. Unless so indicated, such
presentations should not be construed as necessarily ex-
pressions of the author's personal opinions.

THE NATURAL OCCURRENCE OF ALUMINUM IN FOODS

As is to be expected from its abundance and wide distribution in nature, aluminum is of common occurrence as a natural ingredient in many food products. In occasional instances reported it is possible that contamination with adherent soil was not rigidly excluded and that the methods of analysis employed were not strictly accurate. However, while this may have occurred in individual cases, it has been clearly established that aluminum is a natural ingredient of a wide range of food products.

The following data are cited for the most part from Langworthy and Austen.[1] For the purpose of uniformity, results otherwise expressed have been transposed into the equivalent Al_2O_3, as which they are here reported:

TABLE I

VEGETABLE KINGDOM

A. CEREALS AND VEGETABLES

	Total Ash Per Cent	Al_2O_3 Per Cent Ash	Al_2O_3 Per Cent Whole Product
Barley, air dry	1.09	0.140	
Bean, red, air dry	2.60	0.096	
Bean, soy, air dry (whole)	.....	0.053	
Beets: root, dry	13.4	0.28	
root, dry	11.27	0.25	
leaves, dry	20.86	0.12	
leaves, dry	15.57	1.99	
Buckwheat, air dry	1.72	0.113	
Coffee, beans	.....	Trace–2.78	
Corn, kernel	.....	Trace	
Dandelion, dried plant without root	8.88	0.402	
Fungus, edible (Bovista gigantaea)	0.571	15.66	
Oats, dry	3.06	0.07	
Peas, dry	2.10	0.17	
Peppers, red	6.76	Trace	

TABLE I. (*Continued*)

	Total Ash Per Cent	Al_2O_3 Per Cent Ash	Al_2O_3 Per Cent Whole Product
Potato (sweet)....................	1.51	Trace	
Rice (hill) air dry.................	0.87	0.161	
(paddy) air dry..................	0.56	0.189	
Rye, dry..........................	1.90	0.50	
Tea leaves........................	.....	0.81–2.226	
Truffles, white....................	1.80	6.9	
black...........................	2.09	5.3	
Wheat, 22 samples.................	.....	....	0.001–0.070*
air dry........................	2.62	0.106	
winter.........................	1.62	0.11	
grown on sandy soil..............	.....	None	
Yeast, dry........................	9.73	Trace	

* Egyptian, probably contaminated with clay

B. CEREAL AND VEGETABLE PRODUCTS

	Total Ash, Per Cent	Al_2O_3 Per Cent Ash	Al_2O_3 Per Cent Whole Product
Bread, crumb, No. 2 Crown flour........	....	....	0.005
Cal. coarse flour.....................	....	....	0.004
as purchased, 32 samples, crumb......	....	....	0.001 –.003
from various flour, 15 samples........	....	....	0.001 –.005
Flour			
No. 2, Crown.........................	....	....	0.007 –.010
Various, 33 samples..................	0.26–1.44	....	0.0004–.0085
⌈ Wheat (original English grown on chalk soil)..............................	1.72	....	0.005
∣ Bran (18.5 per cent of wheat).........	5.64	....	0.007
∣ Sharps (8.5 per cent of wheat)........	2.00	....	0.007
∣ One Crown (26 per cent of wheat).....	0.50	....	0.003
⌊ Four Crown (45 per cent of wheat)....	0.368	....	0.003
Various, 12 samples....................	....	....	0.004 –.025
Various, 6 samples....................	....	....	0.023 –.080
⌈ Wheat, winter.......................	1.62	0.11	
∣ Bran..............................	5.25	0.07	
∣ Ship stuff..........................	3.08	0.18	
⟨ Dust room..........................	2.50	0.04	
∣ Low grade..........................	0.70	0.12	
∣ Straight...........................	0.40	0.15	
⌊ Patent............................	0.31	0.41	
Sugar, cane.............................		0.65	

TABLE I. (*Continued*)

C. FRUIT AND FRUIT PRODUCTS

	Total Ash, Per Cent	Al_2O_3 Per Cent Ash	Al_2O_3 Per Cent Whole Product
Almond................	....	0.138	
Apple, ripe............	0.30	0.80	
Blueberries, dry.......	....	....	0.005
Cherry...............	0.516–0.739	0.80 –2.55	
Chestnuts, dry flesh.....	3.02	0.09	
Currants, red and white juice................	....	....	Trace
Grape, red............	....	....	0.0027
Grape juice, 18 various..	0.252–0.395	0.021–1.60	0.004
Orange peel............	1.21	0.069	
Plum, fresh............	0.407	....	0.003
Strawberry, juice......	....	....	Trace
Vinegar, cider.........	....	Trace 0.158–1.654	
Wine, 14 various.......	0.030–0.593	0.377–3.956	
7 various.............			0. –0.0036
7 various.............			0.0004–0.0072

TABLE II

ANIMAL KINGDOM

Beef, fresh............	2.2	Trace	
Egg albumen...........	2.3	6.29	
Honey................	...	.....	0.0046
Veal, breast...........	3.1	Trace	

TABLE III

NATURAL WATERS

A. LAKES

	Al_2O_3 Parts per 100,000
Borax, Calif............	0.29
Gmunden, Austria-Hungary............	0.1
Goodenough, Brit. Columbia............	34.5
Loch Katrine, Scotland............	0.096
Michigan, U. S............	Trace
Mono, Calif............	45.78
Owens, Calif............	2.3
Superior, U. S............	0.09
Yellowstone, U. S............	0.4

TABLE III. (*Continued*)

	Al₂O₃ Parts per 100,000

$$Al_2O_3$$
Parts per 100,000

B. RIVERS

Belabula, Clifden Run, N. S. W.	Trace
Detroit, Mich.	1.05
Doubs, at Rivotte.	0.0021
Humbolt, Nev.	0.13
James, Va.	0.0071
Loire, at Orleans.	0.0071
Mississippi, at Carrollton.	3
Mississippi.	3.13
Neponset, at Hyde Park, Mass.	0.0062
Ottawa, near Montreal.	Trace
Parana, above La Plata.	0.3
Passaic, N. J.	1.342
Rhine	
at Cologne	0.0025
at Strassburg	0.1
Rhone, at Geneva.	0.0039
Rio Grande.	0.21
Rio de la Plata, above Buenos Ayres.	0.6
Rio Vinagre.	40
St. Lawrence, near Vandreuil.	Trace
Seine, at Bercy.	0.005
Tarrgagar, near Dubbo, N. S. W.	Trace
Thames, Eng.	Trace

C. SPRINGS, POTABLE

Apollinaris, Germany.	1.2
Atchison Parker, Kans.	0.32
Stafford, Conn.	0.11
Trinkquelle, Driburg.	0.0106
White Rock, Wis.	1.07

D. WELLS, POTABLE

Auditorium, Chicago, Artesian well.	0.12
Emperador, Panama.	2.0
Geneseo, Ill., Artesian Well.	14.67
Merrick, L. I.	8.0

E. MINERAL SPRINGS AND WELLS

Acid Springs, Iowa.	338.35
Acid water, Lake White Island, New Zealand.	931.4
Alum Spring, Rockbridge Co., Va.	1241.4
Alum-water, Lee Co., Va.	89.1

TABLE III. (*Continued*)

	Al_2O_3 Parts per 100,000
Bethesda Springs, Wis.	0.21
Blue Lick, Mo.	17.54
Buffalo Lithia, No. 2, Va.	3.87
Burr Oak, Kan.	11.75
Celestine, Vichy.	1.3
Civillina, Italy.	45.01
Elizabethenquelle, Baden.	.07
Geneva Lithia, N. Y.	4.48
Grande Grille, Vichy.	18.0
Hot Springs, Ark.	0.77
Hot Springs, Nev.	0.19
Hunter's Hot Springs, Mont.	0.23
Jordan Alum Springs, No. 4, Va.	41.50
Kantavu, Fiji Islands.	9.09
Karlsbad Sprudel.	2.8
Kentucky, Alum Springs.	20.61
Kopuowhara Makia, New Zealand.	9.20
Londonderry Lithia, N. H.	2.59
Mt. Clemens Mineral Well, Mich.	50.55
Oak Orchard Acid No. 1, N. Y.	6.50
Old Alum Well, Harrowgate, Eng.	127.8
Old Faithful Geyser, Yellowstone Park.	.17
Orchard Alum, Eng.	9.20
Overall, No. 1, Tex.	41.83
Rockbridge Alum Spring No. 3, Va.	75.40
St. Clair, Mich.	1419.3
Saratoga Springs, N. Y., Congress.	0.55
Hathorn.	0.22
High Rock.	2.10
Pavilion.	0.72
Shenandoah Alum Springs, Va.	17.84
Sour Lake Mineral Springs No. 7, Tex.	23.31
Tuscarora Sour, Brantford, Canada.	46.81

TABLE IV

MISCELLANEOUS

	Al_2O_3 Per Cent
Edible earth, Fiji.	41.53
Japan.	13.61
Java.	25.94
Lapland.	40.80
(Clay) New Zealand.	17.97

SOLUBLE ALUMINUM COMPOUNDS IN CERTAIN VEGETABLE PRODUCTS

Of the more recent data, particularly informative is the report[2] by C. N. Myers, expert in organic chemistry, and Carl Voegtlin, professor of pharmacology, Hygienic Laboratory, U. S. Public Health Service:

"In order to make the study as complete as possible, samples of corn from nearly every state in the United States were examined, so that a fair average might be expected from a series of chemical analyses. In addition, as many samples of foreign corn as were available were analyzed. It should be stated that all of this corn was of good quality with a high percentage of germination. The percentage of metallic aluminum in this set of about 75 samples varied from 0.074 to 0.181 for whole corn grown in America. Corn in the form of meal, grits, hominy, etc., showed different values. These differences depend largely upon the method of milling employed, and upon what part of the kernel has been removed.

"The products analyzed can be easily summed up in the following brief table expressed in per cent of aluminum:

TABLE V

Whole corn (American)	
Maximum	0.181
Mean	.120
Minimum	.074
Whole corn (foreign)	.165
Corn oil cake	.817
Corn meal	.178
Hominy	.145
Grits	.168
Oatmeal	.175
Millet	.428
Rye	.273
Wheat flour	.045
Sunflower seed (hull removed)	.401
Cottonseed meal	.996
Banana (fresh, peel removed)	.033
Yams	
Dried	.332
Fresh	.094

TABLE V. (*Continued*)

Parsnips
    Dried.. .306
    Fresh.. .053
Carrots
    Dried.. .301
    Fresh.. .036
Irish potatoes
    Dried.. .126
    Fresh.. .029

"In the above table it is noticed that cottonseed meal has a very high percentage of aluminum, and wheat flour the least (for the dried material). In the case of the vegetables, the calculations were made upon the basis of both fresh and dried vegetables. Yams contain about 72 per cent of water, and carrots 88 per cent; thus it is seen that very large amounts of aluminum are present when calculations are made for the dried vegetables.

"The question then arose, is the aluminum present in a water-soluble form in these foodstuffs? It is obvious that only soluble aluminum may be absorbed from the gastrointestinal canal. In order to investigate this phase of the subject, a great many water extractions of the vegetables concerned were made, and the filtered extracts analyzed for aluminum. Among the foodstuffs examined were white potatoes, boiled in a manner similar to the method used by the housewife, sweet potatoes, carrots, parsnips, oatmeal, hominy, corn meal, etc. A second series of experiments was carried out by digestion with distilled water in a thermostat for twelve hours at 37°C. A third series of experiments was conducted with the same material in the presence of 0.3 per cent hydrochloric acid (same as gastric acidity) in a thermostat for twelve hours, at 37°C. In order to study the effect of baking on the solubility of the aluminum contained in corn, bread was made from corn meal which had previously been analyzed for aluminum. This bread was then submitted to water and weak hydrochloric acid extraction.

"As a result of these four series of experiments, it is found that at least 65 per cent of aluminum present is soluble in water at 37°C. In the presence of hydrochloric acid (a condition prevailing in the stomach), the aluminum of corn, carrots, etc., is completely soluble. On the other hand, wheat flour and sunflower seed contains but a trace of soluble aluminum. In the case of one sample of corn which was previously analyzed, all of the aluminum present was soluble in weak hydrochloric acid. The following table shows examples of all series of experiments previously outlined:

TABLE VI

	Total amount of aluminum in dry vegetables, per cent	Aluminum soluble in H_2O, digested 12 hrs., 37°C., per cent	Aluminum soluble in 0.3 per cent HCl digested 12 hrs., 37°C., per cent
Sweet potatoes...............	0.332	.....	0.261
Cottonseed meal.............	.996	.....	.718
Millet......................	.428	.....	.171
Corn (whole)................	.115	0.090	.115
Carrots.....................	.305	.....	.312*

*The carrots used in this determination were a little withered, and undoubtedly the amount of dry material was greater than that found in previous water determinations.

"The corn meal which was used for the preparation of the corn bread contained 0.217 per cent of aluminum. Equal amounts of this meal were baked to various degrees of brownness (light, medium and dark).

"A sample of light brown bread yielded on extraction with water for twelve hours at 37°C., 0.169 per cent aluminum.

"A sample of medium brown bread yielded on extraction with 0.3 per cent HCl for twelve hours, at 37°C., 0.197 per cent aluminum.

"A sample of dark brown bread yielded on extraction with 0.3 per cent HCl for twelve hours at 37°C., 0.201 per cent aluminum.

DISCUSSION AND CONCLUSIONS

"The presence of aluminum in cereals and in vegetables has received little attention from investigators who had occasion to study the inorganic constituents in the ash of these products. Reference is made here to the bibliography by Langworthy and Austen on the 'Occurrence of Aluminum in Vegetable Products, Animal Products, and Natural Waters.' From an examination of the data given in this paper it will be seen that certain cereals and vegetables which are consumed in relatively large quantities by man, contain considerable amounts of soluble aluminum compounds (corn, sweet potatoes, carrots, etc.). On the other hand, some cereals such as wheat, for instance, which are also consumed in relatively large amounts, contain aluminum in a mainly insoluble form . . .

"In conclusion, it may be said that aluminum is present in certain foods in rather large quantities in a water-soluble form, and that the daily consumption of aluminum on a mainly vegetable diet may assume larger proportions."

Recently, Philip R. Gray has reported[3] determinations using the modified Atack colorimetric method (p. 219), of the amounts of aluminum in certain products, in most instances as purchased on the market, as follows (parts of aluminum per million):

TABLE VII

Milk........................	1.5 and 2.4
Butter.......................	4.2
Egg white....................	0.2 and 0.4
Egg yolk.....................	2.2, 2.8, 3.5, 3.6, 4.3, 5.6, and 11.2
New potatoes.................	3.3
No. 1 hard spring wheat, whole grain.......................	7.4 and 5.9 (duplicates), average, 6.65
Yellow corn meal, Lily White brand......................	9.7 and 11.2
White corn meal, unbolted, Cornetia brand..................	9.4

REASONS FOR PRESENCE OF ALUMINUM AND ITS FORM

Considerable interest has attached to the reason for the presence of aluminum in plants and plant material and to the form in which it is present. That in some instances, and presumably not infrequently, it is present in part merely as a result of contamination with soil is obviously true. Thus, there is undoubtedly force in the contention of Penney[4] that part of the alumina of certain Egyptian wheat and flour was due to contamination with clay through careless packing in Nile boats. However, the contention of Blyth[5] that properly cleansed wheat contains no alumina, such as is present being always in the form of a silicate and attributable to clay contamination or to sand derived from millstones when the wheat is ground, is at variance with observations of others that must be given weight. Particular credence must be given to Teller's careful investigation.[6] He not only employed iron rollers for crushing the wheat, but found the same amount of alumina in the uncrushed wheat that had been subjected to most careful cleansing with distilled water to remove any possible adhering clay. On the other hand, wheat specially grown on sandy soil contained no alumina. Young's investigations[7] give further basis for the conclusion that, as ordinarily grown, alumina is present in wheat as a natural constituent.*

In a study of the aluminum in the ash of different portions of several plants, Ricciardi[8] concludes that, generally speaking, it is most abundant in the trunk and branches, less so in the husks and seeds and least of all in the leaves, and that assimilation of alumina does not depend on the percentage in the soil.

We have no knowledge of the form in which mere traces of alumina exist in plant structures. Church[9] is authority for the statement that in club moss (Lycopodium) in some samples of which alumina may make up as much as one-third of the total ash, an abundance of an organic aluminum salt may be extracted with boiling water. This has been variously held to be an acetate (John,[10] Mayer[11]), malate (Ritthausen[12]), tartrate (Arosenius[13]) and oxalate (Knop[14]). An organic combination was definitely indicated (Maiden and Smith[15]) by the finding in the interior of a silky oak log (Crevillea robusta, Orites excelsa) of a large amount of a white substance having the appearance of whiting, which proved to be basic aluminum succinate. The investigators regard it as probable that the aluminum was originally present in the sap of the tree as a malate and that the deposit was the result of a natural effort to get rid of a surplus of aluminum not needed by the tree. In Smith's later investigation,[16] the composition of the tree ash indicated about 40 per cent of alumina, much of

*See also pp. 176–181.

which was soluble in water as a potassium salt. Carbonates were not present, because of which "it is supposed that the potassium aluminate was originally present in the tree as such." Yoshida[17] found alumina in the Japanese lacquer tree (Rhus vernicefera), apparently as an arabate. Young in examining wheat flours found the alumina associated and apparently combined with the gluten, and not present in the starch. This is at least consistent with the distribution in the various wheat products.

The considerable interest attached to the presence of aluminum in certain foods, notably wheat and grapes and their products, is owing to the addition of alum as an adulterant to obscure inferiority and whether or not the presence of small amounts indicates such additions. The occurrence in natural waters is of interest chiefly as showing the wide distribution and the ingestion of considerable amounts of soluble aluminum salts by the drinking of such water.

NATURAL ALUMINUM CONTENT OF A SEVEN DAYS' DIET

Basing his calculations chiefly upon the foregoing data, B. R. Jacobs, Chemist, of Washington, D. C., specializing on the chemistry of cereal food products, has recently estimated[3] the natural aluminum content of a seven days' diet for one adult person as follows:

TABLE VIII

DAY I

Breakfast

Ounces		Grams, as $AlPO_4$ in Whole Product
4.6	1 baked apple	0.0081
5.4	3 corn meal pancakes	1.0941
	(with honey) sugar	
	flour	
4.2	2 soft boiled eggs	.1900
8	2 cups coffee	
	sugar	
	cream	
3	Corn bread	.4806

TABLE VIII. (*Continued*)

Ounces		Grams as, AlPO₄ in Whole Product

Lunch or Supper

Ounces		Grams as, $AlPO_4$ in Whole Product
6.3	3 stuffed eggs..	.2850
1.4	2 slices rye bread.................................	.0029
	banana salad....................................	.2100
5.5	(with) bananas................................	
0.06	almonds................................	
1	cooked salad dressing (flour, vinegar and sugar)	
5.2	1 slice plum cake.................................	
	plums................................	.0048
	sugar................................	
	30 gm. white flour.............................	.0060
	½ egg....................................	.0470
8	2 cups tea....................................	
	with 3 teaspoonfuls sugar....................	
	or	
	Apollinaris water...........................	.0048

Dinner

Ounces		Grams
3.2	2 thick slices roast beef...........................	trace
1	with 1 helping Yorkshire pudding................	
	⅓ egg..	.0320
	30 gm. flour....................................	.0060
3.4	1 serving scalloped corn...........................	trace
	½ egg..	.0470
	flour..	
5.8	1 serving fresh parsnips	
	or	
	1 serving dried parsnips...........................	.4149
14	2 baked yams..	1.7760
3	egg and tomato salad (1 tomato, 1 egg)............	.0950
	with mayonnaise (1 egg).......................	.0950
3.5	1 serving grape Bavarian cream....................	
	grape juice.............................	.0154
	egg whites.............................	.1900
	1500 cc. water................................	.0240
	Total for day....................................	5.0286

Table VIII. (*Continued*)

DAY 2

Breakfast

Ounces		Grams, as AlPO$_4$ in Whole Product
7	1 glass grape juice	.0154
6	1 dish cream of barley	.0012
0.75	sugar	
4.2	2 soft boiled eggs	.1900
2.2	3 bran muffins with honey	.0009
	20 gm. wheat flour	.0040
	coffee	
8	cream	
	sugar	
3	corn bread	.4806

Lunch or Supper

1.7	1 serving cheese soufflé	
	¾ egg	.0720
	flour	
5	3 blueberry muffins	
	blueberries	
	flour	.0134
	sugar	
	¼ egg	.0240
6.9	1½ steamed apples	.0122

Dinner

4	2 thick slices breast of veal	trace
	with 1 large serving bread stuffing	
	¼ lb. bread crumbs	.0125
	⅖ egg	.0260
2.5	1 serving carrot salad	.1154
4	2 servings peas	.0028
5.3	1 large baked potato	.1998
8	1 serving rice custard	
	rice	.0025
	1 egg	.0950
4	1 cup black coffee	
	or	
	Apollinaris water	.0048
	1500 cc. water	.0240
		————
	Total for day	1.2965

TABLE VIII. (*Continued*)

DAY 3

Breakfast

		Grams, as AlPO$_4$ in Whole Product
Ounces		
4.6	1 stewed apple	.0081
	sugar	
6.8	1 serving hominy grits	.4500
5.6	4 slices French toast	.0012
	½ egg	.0475
	sugar	
8	2 cups coffee	
	sugar	
3	corn bread	.4806

Lunch or Supper

8.1	2 servings omelet, 2 eggs	.1900
	with rice	.0019
	or	
	peas	
2.6	4 bread sticks, 70 gm	.0140
	with 3 tablespoonfuls honey	.0009
3.5	1 serving grape jello	
	grape juice	.0154
	sugar	

Dinner

5.4	2 servings baked beans	.0100
6	apple and almond salad	
	(with) apple	.0081
	almonds	
	mayonnaise (containing ½ egg, vinegar)	.0475
7.7	2 servings buttered beets	.1466
2	1 serving tipsy pudding	
	½ egg	.0475
	sponge cake (containing 15 oz. flour, ½ egg)	.0500
	57 gm. wine sauce	.0100
4	1 cup black coffee	
	1500 cc. water	.0240
	Total for day	1.5533

TABLE VIII. (*Continued*)
DAY 4
Breakfast

Ounces		Grams, as AlPO₄ in Whole Product
9	1 serving preserved strawberries.....................	trace
4	1 serving steamed rice.............................	.0025
	with sugar.......................................	
	with cream......................................	
	3½ lb. bread crumb pan cakes....................	
6	white flour, ½ cup.............................	.0060
	bread crumbs..................................	.0250
	sugar ½ tablespoonful.........................	
	½ egg..	.0475
3.3	with honey....................................	.0009
8	2 cups coffee....................................	
	with sugar....................................	
	and cream....................................	
3	corn bread.....................................	.4806

Luncheon or Supper

	creamed dried beef................................	trace
	milk..	
2.4	½ tablespoonful flour..........................	.0010
	butter...	
1.4	2 slices of toast................................	.0078
1.7	stewed rhubarb................................	
	sugar..	
4	tea..	
	with sugar....................................	

Dinner

8	½ lb. beef tenderloin.............................	trace
	½ lb. scalloped potatoes.........................	.1998
8	milk ½ cup....................................	
	10 gm. flour 1 tablespoonful...................	.0020
1.2	¼ cup stewed peas..............................	.0020
6.3	(2) prickly pear salad...........................	
	(6.3 oz.) with mayonnaise......................	
1.0	vinegar.......................................	
	½ egg..	.0475
6	cherry pudding.................................	
	1.2 oz. egg...................................	.0475
	⅛ lb. flour..................................	.0114
	⅙ lb. sugar..................................	
	cherries 2.3 oz................................	.0042
	1500 cc. water................................	.0240
	Total for day.................................	0.9097

TABLE VIII. (*Continued*)

DAY 5

Breakfast

Ounces		Grams, as AlPO$_4$ in Whole Product
2.4	stewed blueberries..................................	.0048
6.0	(1 tablespoonful) corn meal mush.................	.0801 .
	with grape jelly....................................	
.8	1 cup grape juice..............................	.0154
	sugar..	
5.4	2 poached eggs.........................	.1900
4	coffee.............................	
	with sugar..	
	and cream......................................	
3	corn bread.........................	.4806

Lunch or Supper

8.0	(1 orange) orange salad...........................	.0121
	with mayonnaise	
	50 gm. candied orange peal......................	
	egg, ½...	.0475
	vinegar.....................................	
	pepper jam sandwiches...........................	
9.0	red peppers 6 oz...............................	
	sugar, 6 oz...................................	
	vinegar, 2 tablespoonfuls........................	trace
5.0	cocoa......................................	
	milk 1 cup..................................	
	sugar 2 tablespoonfuls.........................	

Dinner

2.0	beef heart (stuffed)................................	trace
14.0	baked sweet potatoes...........................	1.7760
	(Yams)	
5.8	parsnips (browned in butter)......................	.4149
	grape salad....................................	.0007
3.0	with mayonnaise	
	½ egg...	.0475
	vinegar.......................................	trace
2.5	currant tarts.................................	
	1500 cc. water.................................	.0240
		———————
	Total for day.................................	3.0936

TABLE VIII. (*Continued*)
DAY 6
Breakfast

Ounces		Grams, as AlPO₄ in Whole Product

Ounces		Grams, as AlPO₄ in Whole Product
9.5	(3) mandarin oranges	
6.0	cooked Farina ¾ cup	.0144
0.5	1 teaspoonful sugar	
1.8	cream	.
8.4	veal kidneys, stewed	trace
1.4	on toast (2 slices)	.0078
8.0	coffee (2 cups)	
	with sugar	
	and cream	
3.0	corn bread	.4806

Luncheon or Supper

12.0	Turkish rice (tablespoonful rice)	.0014
	beef, ¼ lb	
	red peppers 2 oz	
	tomatoes 4 oz	
2.6	bread rolls	.0056
5.0	(1 apple) apple salad	.0081
	with mayonnaise	.0475
4.0	tea	
	with sugar	

Dinner

	2 creamed eggs	.1900
12.0	milk ½ cup	
	10 gm. flour 1 tablespoonful	.0020
	butter 1 tablepoonful	
6	(2 medium) brown potatoes	.2268
10.5	carrots and	.1359
	peas	
	in butter	
	(⅛ lb.) grape	.0024
8	and banana salad	
	mayonnaise	
1.0	½ egg	.0475
	vinegar	
	chocolate blanc mange	
4.5	milk	
	cocoa	
	sugar	
	egg ½	.0475
2.0	rye bread	
	1500 cc. water	.0240
	Total for day	1.2415

TABLE VIII. (*Continued*)
DAY 7
Breakfast

Ounces		Grams, as AlPO₄ in Whole Product

Ounces		Grams, as AlPO$_4$ in Whole Product
8	(2) orange juice...............................	
6	(⅔ cup) corn meal gruel.........................	.5167
	with sugar.................................	
	and butter................................	
4.2	(2) scrambled eggs............................	.1900
1.4	on toast (2 slices)............................	.0006
4.0	coffee.......................................	
	with sugar.................................	
	and cream.................................	
3.0	corn bread..................................	.4806

Luncheon or Supper

3.0	(½ cup) fruit salad, including 2 teaspoonfuls dressing.	
	28 gm. grapes...........................	.0002
	1 apple.................................	.0081
	almonds................................	
	½ egg..................................	
	mayonnaise ½ tablespoonful.............	.0475
8.0	Parker house rolls............................	.0100
3.3	honey (3 tablespoonfuls)......................	.0009
5.0	milk (1 cup).................................	

Dinner

4.6	veal steak...................................	trace
7.2	(2 medium) mashed potatoes...................	.2268
9.2	(2) baked apples..............................	.0162
	sugar (2 tablespoonfuls)...................	
	butter (2 tablespoonfuls)..................	
	20 gm. flour (2 tablespoonfuls).............	0.0034
10.0	(½ cup) stewed carrots.......................	.1980
	orange...................................	
	and green pepper salad....................	
	½ orange................................	
	¹⁄₁₆ green pepper.........................	
	½ egg...................................	
	2 tablespoonfuls mayonnaise..............	.0475
	1500 cc. water............................	.0240

Total for day...............................	1.7705
Grand Total for 7 days.......................	14.8937 gm.
Equivalent Al...............................	3.3018
Equivalent R.B. units, on the assumption of 70 kg. body weight (see p. 326)......................	6.74

III

ADDED ALUMINUM COMPOUNDS IN FOOD

Aluminum compounds are added to pickles and similar products in the form of alum and to baking powders in the form of sodium aluminum sulphate (s.a.s.). In each of these instances they serve a useful purpose. Besides such additions, a very small amount of aluminum may enter the food from the use of cooking utensils and in occasional instances small quantities may remain in municipal water supplies as a result of the use of aluminum sulphate for purification purposes.

ALUM IN PICKLES

Alum is added to cucumber pickles and to maraschino cherries to preserve their firmness. In the case of pickles, it originated as a domestic practice. In an examination by the author, cucumber pickles processed without the addition of aluminum compounds did not show the presence of aluminum in weighable amount in the quantity of material employed for analysis, though sulphate, as SO_4, was present in the liquor to the amount of 0.03 per cent. On the other hand, in instances where aluminum compounds were added in the regular practice of processing, the liquor showed:

TABLE IX

Samples, 100 gm.	Al gm.	SO_4 gm.	Ratio of Al to SO_4
1. Sweet gherkins.....................	.00149	.0288	1 : 19.3
2. Sour gherkins.....................	.00240	.0457	1 : 19.0
3. Medium-sized pickles...............	.00908	.1268	1 : 14
4. Medium-sized pickles...............	.00278	.0746	1 : 26.9

The form of the aluminum compound used in Samples 1 and 2 was unknown, in Sample 3, alum (presumably the potassium), and in Sample 4, aluminum sulphate. Since the ratio of aluminum (Al) to sulphate (SO_4) is $1:7.1$ in alum and $1:5.3$ in aluminum sulphate, it is apparent that an excess of sulphate was present in the liquor and that the aluminum-sulphate ratio afforded no evidence as to the form in which the aluminum was added.

Samples of the pickles were rinsed free from adherent liquor and the retained liquor largely removed by pressure. The pulp-cakes thus obtained showed in Sample 2 an $Al:SO_4$ ratio of $1:1.6$; in Sample 3, of $1:0.86$; and in Sample 4, of $1:1.02$. In the case of Sample 2, the pulp-cake was further washed until practically acid-free, when the $Al:SO_4$ ratio became $1:1.3$. It is thus apparent that the aluminum retained by the pulp-cake was chiefly in some other combination than alum or aluminum sulphate. It is not indicated whether or not it was wholly in some other form.

Analysis of samples removed from the liquor, which were thus in the same condition as ordinarily eaten, showed:

TABLE X

Samples	Al per cent	SO₄ per cent	Ratio of Al to SO₄
2	0.0123	0.0601	1:5
3	0.0255	0.101	1:4
	0.0406	0.1856	1:4.6
4	0.0101	0.0589	1:5.8
	0.0171	0.0917	1:5.4

While the relative amount of sulphate present was such as to render it possible that the aluminum present was largely in the form of the sulphate, the previous findings in the case of the pulp-cake would indicate that such was not the case. As further evidence upon this point, the pickle juice (so termed to distinguish it from the surrounding liquor) of Sample 4, obtained by expressing the retained liquor, was submitted to dialysis through parchment paper. The dialysate contained, in two such tests, Al and SO_4 in the ratio $1:10$ and $1:11.9$ respectively, thus indicating that the sulphate present was not wholly in combination with aluminum but was in part in some other combination. On the other hand, it indicated that a small amount of aluminum was present in a soluble and dialyzable form.

From these observations the conclusion is indicated that aluminum, added to cucumber pickles in processing, either as alum or aluminum sulphate, to render the product firm, is present in the final product in two forms, one insoluble and apparently in combination with the natural food ingredients; another form soluble and dialyzable, quite likely in the same combination as was originally added. So far as the limited examinations recorded give evidence, it would appear that the greater part was in the insoluble form.

From these data, it appears that when pickle is consumed to the amount of 15 gm. (½ oz.), which is probably as much as would ordinarily be consumed at one time and perhaps in excess of the average, the amount of aluminum ingested would be 1.5 to 6. mg. In drinking Apollinaris water, the aluminum ingested per glass (¼ l.), as calculated from the figures previously given, is 2.25 mg., although in the latter instance it is entirely in solution.

We know of no studies of aluminum in maraschino cherries, but it is fair to believe that the added alum or aluminum sulphate bears the same relation to the food product that we have noted in the case of cucumber pickles. It is also probable that the retained aluminum is in substantially the same amount and in greater part combined and insoluble, with a lesser quantity present in a soluble and dialyzable combination.

ALUMINUM FROM CONTAINERS AND COOKING UTENSILS

The general use of aluminum cooking utensils began about 1890. Lübbert and Roscher,[1] using shavings of aluminum, concluded that aluminum was attacked by wine, coffee, tea, etc. This investigation was followed in 1892 by exact studies on the subject by Lunge and Schmid,[2] who suspended aluminum strips in various solutions for periods of six days, calculating the loss in weight as milligrams per 100 sq. cm., with the following results: tea, and beer 0.0, coffee 0.5, 50 per cent alcohol 0.61, brandy 1.08, red wine 2.84, white wine 3.27, various food acids in 1 to 5 per cent solutions 1.08 to 4.77. They reached the conclusion that the amounts of aluminum dissolved were too small to render harmful the use of aluminum flasks, etc. Rupp,[3] allowing contact for from four to twenty-eight days, likewise reached this conclusion. Balland,[4] extending the time of contact to four months, reached a similar conclusion.

In 1893, Ohlmüller and Heise[5] found that 500 cc. of distilled water reduced the weight of aluminum vessels 1.12 mg. and 3.79 mg. in two and six days, respectively. In boiling, the amount was scarcely demonstrable. With city water, however, the vessels lost 11 mg. to 13 mg. in one-half hour's boiling. It attacked clean metal ten times as strongly as unclean metal. With 500 cc. of 1 per cent acetic acid, the loss in two days averaged 5 mg., in one-half hour's boiling with 400 cc., 52 mg. and 59 mg. and there was also some alteration on the inner side of the vessel.

In the same year, Plagge and Lebbin[6] published the results of investigations on the use of canteen and cooking vessels of aluminum. Water, coffee, vinegar, beer, wine, cognac or other spirits, or lemonade, underwent no change of taste when stored in aluminum canteens, even for periods of a week. Even after sixty days, no change of taste was detected, but after long keeping in some instances the liquid became turbid and spot formation on the vessels occurred. Especially with brandy, even twenty-four hours' keeping with heat brought out brownish spots, attributed to the action of tannic acid. Distilled water did not produce spots but, in the presence of weak acid or salt solution, whitish spots attributed to the action of silicic acid were observed.

Aluminum cooking vessels in general use over a period of one and one-half years gave no suggestion of metallic taste to food. A darkening with boiling natural water was attributed to iron sulphide. Five hundred cc. natural water, either at room or incubator temperature, at rest or shaken, took up only a fraction of a milligram in eight hours; 1 liter of ½ to 2 per cent table salt solution, by boiling twenty-four hours, took up 25 mg. Al_2O_3. Five hundred cc. $\frac{1}{2}$ per cent acetic acid solution, (1) at rest or (2) shaken, in eight hours at room temperature took up (1) 3 mg. and (2) 7 mg. Al_2O_3; at 37°C. (1) 15 mg. and (2) 23 mg. Brandy acted much the same. Wines under similar conditions absorbed from 1 mg. to 8 mg., coffee, about $1\frac{1}{2}$ mg. per 2 liters. The use of aluminum vessels for cooking meat and vegetables was estimated to introduce 1 mg. to 2 mg. Al_2O_3 per person per day.

Donath,[7] in 1895, found no material action on aluminum plates by waters or oils. Franck,[8] in 1897, noted the loss of weight of aluminum utensils over periods of three years, as follows: plates boiled in water three hours daily, 0.0143 gm. per 100 sq. cm.; cooking vessel of 732 sq. cm. surface in contact with boiling milk twice daily for one-fourth hour, 0.5139 gm.

In 1904, Mansfeld[9] concluded that considerable amounts of aluminum went into solution from boiling with 4 per cent acetic acid or with 5 per cent soda solutions, thus limiting the employment of aluminum vessels to neutral reacting fluids.

Von Fillinger,[10] in 1908, studied the action of 200 cc. of boiling fluids when 41.6 sq. cm. aluminum surface was exposed for one-half hour. Fresh and sour milk up to an acidity of 6.6. cc. N in 100 cc. had no action. At an acidity of 9.4 cc. N in 100 cc., the aluminum lost 0.2 mg. There was no action by red or white wine. One-tenth normal solution of NaCl, KI, Na silicate, K_2SO_4, $MgCl_2$, $CaCl_2$, $MgSO_4$, $CaSO_4$ and Ca silicate showed no action; $NaHCO_3$ showed a loss of 88 mg. With a mineral water containing much alkali carbonate the aluminum lost 12 mg.

In 1912, A. Barillè,[11] studying the use of aluminum as material for siphon heads, found that the metal was attacked four times as strongly by the water saturated with carbon dioxide gas as by the ungassed water and that the action was accentuated by gas pressure.

In 1913, The Lancet[12] published the results of extensive experiments conducted in their own laboratory "bearing directly on the question as to whether aluminum is appreciably acted upon by the usual articles of food or by the usual acids employed in the preparation of food when it is treated in the (aluminum) kettle, frying pan, or sauce pan." Their conclusions follow:

"The experiments on the whole show no serious count against aluminum. This metal, at all events, does not appear to be more susceptible to the action of water and foods in the process of cooking than does iron, which has been used from time immemorial as the material of cooking pans. As is well known, iron rusts very readily in the presence of water and air, while also it is attacked by organic acids. It is also well known that iron salts in large quantities are injurious to the organism, as are also large quantities of aluminum salts, as for example, alum (the double sulphate of aluminum and potassium or ammonium), and aluminum chloride. But there is no evidence to show that in the ordinary cooking operations of everyday practice either iron or aluminum is so strongly attacked as to produce an objectionable amount of soluble salts. All that can be found, when even organic acids and mineral salts are present in the cooking pan, are the merest traces of metal in a soluble state. The alumina precipitated by ammonia in the tests was in practically all cases an unweighable quantity. The case is different when an alkali is present. Carbonate of soda, for example, is without action upon iron, but it certainly attacks aluminum freely, and it would be well to exclude that salt from an aluminum cooking utensil, although even in this case it is doubtful whether any injury to health would be done. The makers of aluminum cooking vessels are fully aware of this action, and, as has been said, they commonly issue a notice warning the possessors of aluminum vessels not to use carbonate of soda, which, after all, is not indispensable to cooking processes.

"A considerable number of experiments has been made concerning the action of water upon aluminum chiefly in connection with the suggestion that the metal might be used in the making of the water-bottles and mess-tins of the soldier, with the advantage which its lightness would give compared with all other materials hitherto put to similar use. At first the question of expense was a formidable objection to the idea, and later the experiments were unsatisfactory owing to the impurities present in the aluminum early put upon the market. The first objection has disappeared and with it a purer aluminum containing less than 2 per cent of impurities has been commercially produced. At the present time aluminum is universally used in the manufacture of water-bottles for soldiers and other composite bodies of workers. But certain precautions are necessary in order to keep aluminum vessels suitable and hygienic for the purpose, the chief of which is that water should not be left in the bottle when it is not in use. This is in accordance with the findings of our experiments, for it was shown that there is a tendency for wet aluminum to oxidise when left freely in contact with air. It may be desirable, also, not to scour the vessels to such an extent as to remove any thin coating which forms on the surface, for this coating subsequently becomes protective. When not in use it is obvious that aluminum vessels should be kept perfectly dry and free from dust, and a film of hydrocarbon oil effectually protects the metallic surface against the combined action of moisture and air. The same precautions apply equally to iron pots and pans, but in this case it is generally forced upon the attention of the user, as any neglect to keep the pans clean and dry means an unsightly, reproachful, rusty pan. In the experiments recorded we, as a matter of fact, scoured the pans with silver sand for each experiment, which thus exposed a fresh surface each time so that the conditions were such as to favor a corrosive action, in spite of which no action of serious hygienic importance occurred.

"We are confident that aluminum, as it is now made by reputable manufacturers, is a suitable material for cooking vessels, and that any suspicion that it may communicate poisonous qualities to food in the process of cooking may safely be dismissed in view of the results of the practical experiments which we have recorded, showing that the metal is not appreciably acted upon in cooking operations. This finding is satisfactory also inasmuch as aluminum is an excellent heat conductor; cooking in aluminum vessels is therefore rapid and fuel is economized in consequence. But the management of aluminum cooking utensils requires the same ordinary applications of common sense as are customary in the case of other metals employed for a similar purpose."

In a later issue of the same year, The Lancet[13] referred editorially to the work in similar lines reported by Dr. John

Glaister, Regius Professor of Forensic Medicine and Public Health, Glasgow University, and Dr. Andrew Allison, his senior assistant. In practical cooking experiments, they found no dissolved aluminum with bacon, dripping, or milk, but small amounts with oranges and lemons for marmalade, brussels sprouts and tomato sauce. In laboratory experiments with solutions of known composition, they found no aluminum in the case of common salt, acetic acid, or a mixture of common salt, tartaric acid and citric acid, but minute traces in the case of sodium bicarbonate.

The largest amount found in their experiments was 1.018 grains of aluminum hydroxide in 2½ lb. of marmalade made with oranges and lemons. They concluded that "the ordinary use of aluminum cooking vessels for culinary purposes is not attended with any risk to health of the consumers of food cooked therein." *The Lancet* added: "We welcome this confirmation, which should dismiss any suspicion that with common precautions aluminum is a source of danger as a material of cooking vessels."

In 1914, E. Blough[14] studied the action of tomatoes on various cooking utensils. The increase of aluminum in the food, when allowed to stand in the aluminum utensil twenty-four hours after the cooking, was not greater than the increase of tin in the food removed immediately on the completion of cooking in tin utensils; nor of the increase of iron in the food allowed to stand for twenty-four hours in enamel-ware utensils in which the food was cooked. The increase of aluminum was in most instances no greater when the food was cooked in aluminum utensils than when it was cooked in enamel ware. In all instances the metals derived from the utensils were very minute in quantity.

In 1919, Utz[15] observed that lactic acid (0.5 to 0.801 per cent) had no effect on aluminum at room temperature but at higher temperature dissolved a small amount, which he regarded as harmless. Hence, aluminum vessels were suitable for milk products. Trillat[16] reached the same conclusion.

In 1922, Friend and Vallance[17] concluded that "the presence of protective colloids is probably the explanation of the very marked resistance to corrosion offered by aluminum pans." Thus, studying the action of 3 per cent sodium chloride solution and regarding its mean relative corrosive action as 100, the addition of 0.1 per cent gelatin reduced the action to 44.6.

In 1924, Tinkler and Masters,[18] without wishing to question the correctness of the conclusions reached by *The Lancet*, called attention to their own observations that, for a given sample of aluminum and different samples of natural waters, the intensity of the stain produced by natural waters depended on the hydroxide-ion concentration of the water. An extremely small amount of aluminum was removed from the surface, leaving behind iron and other impurities which constituted the stain. The non-formation of a stain when food was boiled in the water was due to protection of aluminum by the presence of colloidal matter.

In 1924, Gies[19] expressed an opinion unfavorable to the use of aluminum cooking utensils.

Subsequently in a critical examination of the subject, Edmunds[20] concluded that the literature was overwhelmingly in favor of the conclusion that the use of aluminum cooking vessels was harmless.

Likewise, recent writers on pharmacology (Cushny,[21] Poulsson,[22] Sollmann[23]) have expressed the conclusion that the quantities of aluminum that might be dissolved from aluminum cooking vessels, even by dilute acids, were too small to be of any importance.

BAKING POWDERS CONTAINING ALUMINUM SALTS

There are two classes of baking powders in which an aluminum salt is employed as an acid ingredient. In the one class the aluminum salt is the sole acid ingredient and the powders are known as "straight"; in the other, the aluminum salt is one of two acid ingredients, the other of which is calcium acid phosphate, and such powders are known as "combination" powders. Any powders other than these two classes that contain added aluminum salts, are of such exceptional occurrence that they need not be considered.

The particular aluminum salt employed is obtained by evaporating a mixed solution of sodium sulphate and aluminum sulphate in substantially molecular proportions and roasting the residue. The resulting product is known as sodium aluminum sulphate or s.a.s. The aluminum of the sodium aluminum sulphate is derived from the naturally occurring bauxite, essentially $Al_2O_3.2 H_2O$, or gibbsite, Al_2O_3. $3H_2O$, found in large deposits. There has been much controversey with reference to the proper name for this product. Its users maintain that the only proper name is sodium aluminum sulphate. Their competitors do not claim that this name is incorrect or untrue, but insist that it is a chemical name which people do not understand. They claim that popular and trade usage of the term "alum" as applied to s.a.s. and a similar usage of the same term for aluminum sulphate has made "alum" the common name for both these substances as well as for alums of the drug store (K_2Al_2 $(SO_4)_4$. $24H_2O$ and $(NH_4)_2 Al_2(SO_4)_4.24 H_2O$). They further contend that the uses of these four substances (all of which they call "alum") are similar and that they have similar chemical reactions.

The users of s.a.s. claim that the use of the name "alum" for several different substances only results in confusion which they seek to avoid; that s.a.s. is used in baking powder solely for its acidic action, which is an entirely different purpose from the use of the true alums or of aluminum sulphate for producing a precipitate of aluminum hydroxide. They claim also that the name sodium aluminum sulphate is true and distinguishes it from other substances; that the name "alum" applied to a water-free substance is a misnomer and that this view is supported not only by chemical textbooks but also by Federal and State Drug Laws and by the rules governing the classification of freight and the denaturing of spirits, all of which deny the name "alum" to s.a.s.

s.a.s. is used exclusively as an ingredient of baking powder. It is recognized as being a different substance from the aluminum compounds used in other food products and the trades.[24] The users of s.a.s. claim that the question here raised is whether we shall continue to call s.a.s. by its distinctive name, sodium aluminum sulphate, and continue to recognize the limited application of the word "alum," as now defined by law and chemical authorities, or yield to a broader definition of the meaning of alum with its attendant confusion.

The essential feature of the sodium aluminum sulphate, which we shall designate "s.a.s.," is that it is a comparatively non-hygroscopic substance and in the presence of moisture is possessed of the property of reacting as an acid with sodium bicarbonate, the alkaline ingredient of baking powder. It thereby liberates carbon dioxide gas, the aerating agent for which the baking powder is employed, according to Equation 1:

$$Na_2SO_4.AL_2(SO_4)_3 + 6NaHCO_3 = 4Na_2SO_4 + 6CO_2 + 2Al(OH)_3$$

The theoretical proportions in this equation are s.a.s. 484.44, sodium bicarbonate 504.06; that is, in order to react without leaving an excess of either ingredient, the amount of s.a.s. required is 0.961 times the quantity of sodium bicarbonate taken; or, to state the fact in a different way, 1 part of s.a.s. has a neutralizing value of 1.04. In the better grade of materials there is little variation from this theoretical figure. Different batches of a good grade of s.a.s. will show variation of only 1 per cent to 2 per cent from theory. In order, however, to incorporate into the baking powder the active ingredients in their exact reacting proportions, it is desirable to control the relative amounts added by exact chemical determination of the reacting amount with each batch of ingredients used. This is actually done in the manufacture of the better grades of powder. The practice is to have present a slight excess, usually 1 per cent sodium bicarbonate.

"Straight" S.A.S. Baking Powders. The acid ingredient, the sodium bicarbonate and the filler, usually corn starch, with or without dried white of egg, make up the sum total of the constituents of the straight s.a.s. baking powders.* There is a slight reaction between the s.a.s. and sodium bicarbonate during storage of the finished product, due to absorption of atmospheric moisture, which results in some loss of available

* In the following account, we shall recognize as "alum baking powders," those containing a salt of aluminum as an acid ingredient and made in England and in the United States prior to 1893 as "s. a. s. baking powders" those made in the United States subsequent to 1893.

gas and is more marked when subjected to high temperature. As a result of factors such as this, the product as consumed may contain slightly less available CO_2 than what is calculated from the intended composition. This difference is ordinarily small in amount, but when it is sufficient to lower materially the available gas, the product is recognized as inferior and in good practice is withdrawn from the trade. Interaction during storage is less apt to occur with s.a.s. as the acid ingredient than with either acid phosphate or acid tartrate.

"Combination" Baking Powders. Calcium acid phosphate, which together with s.a.s. is an acid ingredient of "combination" powders, was formerly prepared from bone by complete ignition of the organic matter, solution of the ash in dilute sulphuric acid and, after suitable purification, formation of the calcium acid phosphate, which was subsequently dried. Thus prepared, it contained a small amount of calcium sulphate as an impurity. Today calcium acid phosphate is prepared almost exclusively from rock. New methods of manufacture have been introduced which render the phosphate from this source the purest on the market and practically free from fluorine.

Calcium acid phosphate reacts under certain conditions with sodium bicarbonate according to Equation 2:

$$3CaH_4(PO_4)_2.H_2O + 8NaHCO_3 = Ca_3(PO_4)_2 + 4Na_2HPO_4 + 8CO_2 + 11H_2O$$

The residual end-products are tricalcium and disodium phosphates.

The theoretical neutralization value of calcium acid phosphate is 88.9. The actual determinations of the neutralizing value of calcium acid phosphate by titration admits of considerable variation dependent on the conditions of manipulation. Under certain conditions a value as high as 85 may be obtained; a commonly used method yields values of about 82. The addition of sodium bicarbonate to a baking powder in the proportion of even this value would yield a high residual gas (not liberated by water and heat), so that a lower neutralizing value, 72 to 75, is ordinarily used for such a product. The desirable amount of the calcium acid phosphate is such as will yield a small amount only of residual gas. This insures complete reaction of the acid and a very slight excess of the sodium bicarbonate.

In different brands of "combination" baking powders different proportions of the s.a.s. and calcium acid phosphate are employed. In any case of an admixture of the two acid ingredients, the neutralizing value is less than the sum of the neutralizing values of the separate ingredients. This is apparently due to the formation of aluminum phosphate.

Attempts are sometimes made to represent in a single equation the interaction of s.a.s. and calcium acid phosphate when acting together with sodium bicarbonate. That such schemes do not represent what actually occurs in the use of baking powders of this class is apparent when we consider that the times of reaction in the dough of the s.a.s. and acid phosphate are not the same. Indeed upon this fact depends one of the advantages in the use of the "combination" powders. At room temperature, on the making of the dough, the acid phosphate reacts in considerable amount with the sodium bicarbonate, immediately liberating gas for aeration, while, when the dough is warmed in baking, s.a.s. reacts with the sodium bicarbonate to a large extent. In this way, the leavening proceeds over a greater period of time and range of temperature than with a single acid ingredient and the necessity for quick baking is lessened. Also the optimum temperature of the oven for baking is extended over a wider range, so that less skill is required to obtain good results.

Thus any attempt to represent accurately by a simple equation of interaction what actually occurs in the use of a "combination" powder is quite unwarranted. The reactions occur in stages, the earlier being chiefly with the acid phosphate, the intermediate probably with both acid ingredients, and the latter chiefly or wholly with s.a.s.

Moreover, the end-products of interaction between acids and bicarbonate are not limited to the products indicated in Equations 1 and 2. There is reason to believe that there are further formed as end-products one or more aluminum phosphates, which in all probability are of variable, and possibly to some extent quite complex, composition. That with a given "combination" powder the end-products

formed are not invariably of the same composition seems
likewise to follow from the fact that the speed of reactions
varies with the proportion of moisture, the temperature of
the added water and possibly other factors, conditions which
would doubtless change the relative proportions of phosphate
and s.a.s. interacting at a given time. Since chemical reactions between the ingredients of the baking powder must take
place before leavening is effected and the food ready for
consumption, it is the residual products from these chemical
reactions, rather than the original ingredients of the powder,
which are to be considered in their effect on nutrition and
health.

*Interaction between Baking Powder Residue and Food
Ingredients.* Still another factor to be given consideration in
this connection is the possible interaction between baking
powder residue and the food ingredients. That aluminum
hydroxide tends to combine with proteins is well known.
There is intimate physical contact between the baking powder and the food and, further, the chemical reaction is on the
alkaline side of the isoelectric point of the food proteins. On
the other hand, there is a question as to the extent of solubility of the food proteins and the influence of this on
interactions.

With the existence of these complex conditions, it is not
possible with the data at hand to decide the exact nature of
the final baking powder residues in foods. Nevertheless, consideration must be given to the various possibilities. Even
in the absence of phosphate in the baking powder, it is
known, as we shall show later, that the residue from s.a.s.
which in its simplest form is aluminum hydroxide, tends to
deflect phosphate elimination from the urine to the feces.
It is fair to presume, therefore, that at some point subsequent to the hydroxide formation the aluminum is in part at
least(to what extent we do not at present know), combined
with phosphate and is eliminated in the stools in such combination. This illustrates the complexity of interactions into
which the aluminum may enter.

A straight baking powder, containing 30.5 per cent s.a.s.,
would in theory require 31.7 per cent sodium bicarbonate for

neutralization, leaving 37.8 per cent as the amount of starch, etc. Such a powder would produce 16.6 per cent CO_2. The residues of the reaction would be 10.1 per cent as $Al(OH)_3$; 35.6 per cent as Na_2SO_4; and 37.8 per cent starch.

If a rounded teaspoonful, or 7.5 gm., of such a powder were used to leaven 1 qt. of flour, the finished biscuit, while variable, might weigh 545 gm., which would then contain baking powder residue, per 100 gm., of 0.139 gm. as $Al(OH)_3$ (or 0.048 gm. figured as Al), and 0.49 gm. Na_2SO_4; or, per ounce, 0.56 grain as $Al(OH)_3$ (or 0.20 grain figured as Al) and 2.1 grains Na_2SO_4.

Variation in Compositions of Combination Powders. The composition of the combination powders is subject to considerable variation, owing to differences in the relative amounts of the acid ingredients, namely, calcium acid phosphate and s.a.s. In 1901, R. B. Davis stated before the Committee on Public Health of the Massachusetts Legislature that the residue from one such powder consisted, per pound powder, of CO_2 2.19 oz., alumina 0.56 oz., sodium sulphate 3.13 oz., phosphate of soda 1.19 oz., phosphate of calcium 0.72 oz. Using two rounded teaspoonfuls or 15 gm. of such a powder to leaven 1 qt. of flour, biscuits (of previous paragraph) would contain baking powder residue, per 100 gm., of 0.147 gm. as $Al(OH)_3$ (or 0.051 gm. figured as Al), 0.54 gm. as Na_2SO_4, 0.20 gm. as phosphate of sodium, and 0.16 gm. as phosphate of calcium; or, per oz., 0.64 grain as $Al(OH)_3$ (or 0.22 grain figured as Al), 2.4. grains as Na_2SO_4, 0.87 grain as phosphate of sodium and 0.70 grain as phosphate of calcium. With neither the straight nor the combination powders is any attempt made to represent the exact combinations of the constituents of the residues, owing to the previously described complications.

A person eating a half pound daily of such a bread would ingest a maximum of 0.33 gm. as aluminum oxide (0.115 gm. figured as Al) from the baking powder residue. The amount would be increased or diminished according as more or less of the baking powder was used or as a greater or lesser quantity of the food was eaten. However, it would seem most unusual for more than 150 mg. (figured as Al), to be thus consumed; while in many and probably in the great majority of instances the quantity would be less than one-half this amount. On this basis, the ingestion, as baking powder residue, for an entire week would vary from very small amounts up to 1 gm., figured as Al. We have elsewhere seen that as a natural food ingredient the amount may readily be as much as 3.30 gm. in a week. If corn bread and yams were consumed daily and in liberal though not excessive amounts, the ingestion as a natural food ingredient might be considerably higher than this.

ALUMINUM INGESTED IN BAKING POWDER FOOD

B. R. Jacobs[25] has calculated that in the case of biscuits of ordinary size made with a 20 per cent s.a.s. combination baking powder, with the consumption of three per meal for a week there would be ingested in the baking powder residue an amount of aluminum one-sixth as great as that calculated for the natural dietary without baking powder residue, to which reference has been made (pp. 14–21). In short, while aluminum has been spoken of as a foreign ingredient of food, it is in reality a commonly occurring natural ingredient and occurs in amounts that in many instances are unquestionably quite small but in certain combinations of foods are capable of exceeding by a considerable margin what would be regarded as a large ingestion as baking powder residue. Furthermore, at least in the case of corn and potable waters, the water solubility of the naturally occurring aluminum compounds is greater than what we have any reason to believe is the water solubility of the aluminum compounds of the baking powder residue.

IV

HISTORY AND RESEARCH 1828-1897

THE TERM TOXIC OR INJURIOUS AGENTS

In considering the action of a substance as a toxic agent, it is important to have clearly in mind both the meaning and the limitations of the term "toxic or injurious effects." A substance is injurious to an organism when it is inimical to the continuity of the normal structure or the normal processes that constitute the functional activities and life of that organism.

This applies whether the organism be plant or animal, or whether it be a simple structure or a complex structure such for example as constitutes higher animals and even man. As a rule the simpler forms of life are more susceptible to external influences than are those more highly developed. They are more easily injured by external influences and agents, and in general, are short-lived. The higher forms of life not only are more diversified in structure and function but, by reason thereof, are better able to resist external influences and agents, and are longer lived. It is, therefore, to be expected, in keeping with the absence of the protective mechanism possessed by the higher forms of life, that simple forms will be unfavorably affected by influences or agents more frequently and to a greater extent than will the more highly developed forms. While this is the general rule, in an individual case it may not be true. If a given substance is inimical to simpler living structures, it may act similarly upon more complex structures, depending upon whether or not the more complex mechanism of the higher form includes protection from the substance in question. Hence, a substance may be harmful to unprotected lower forms of life and wholly harmless to the complicated and protected higher forms. This is undoubtedly true of a large number and variety of substances, some of which are of unquestioned value to higher forms of life.

The inimical effects are equally concerned with the quantitative relation. What is harmless or even beneficial in small quantities may become injurious when the quantity is markedly increased. In considering the action of any substance on a particular plant or animal organism, simple or complex, these factors must be kept clearly in mind and final conclusions reached only when all the facts are established, so that the part played by the structure of the organism under consideration may be properly evaluated.

For example, if a specimen of simple-cell animal life is placed in a 5 per cent solution of common table salt, it will be injuriously affected by reason of the concentration. Also, injurious results will follow if such a solution, or even a relatively weak solution, be directly introduced into the interior of such simple animal cell. These facts, however, establish no conclusion as to the effects of the ingestion of salt by higher animals or by man. They do, nevertheless, raise questions to be investigated and so may properly be included in the broader study of the effects of salt upon living matter, even though they do not prove it injurious under ordinary conditions to higher animals and to man.

In reviewing the early studies of aluminum compounds as possible toxic agents, we may well begin with the historic case of Madam B.

On September 22, 1828, Madam B., who had previously suffered from continuous vomiting, and who at the time was under the medical care of Dr. Fournier-Des-Champs, drank, in mistake for gum acacia, two or three swallows of a solution of 16 gm. of calcined alum dissolved in a liter of lukewarm water. She complained of very severe pains in the mouth, pharynx and stomach, and vomited throughout the day, during which period she ingested 32 glasses of tepid water. The symptoms gradually subsided during the subsequent four days, when she was declared better.

The case came to litigation and her physician maintained that she had suffered severely from alum poisoning, the effects including dilatation of the heart. Two experts jointly investigated the subject of the toxicity of alum. Dr. Devergie[1] held that calcined alum was poisonous, since 60 gm. might produce death of a dog, and even one-half that amount, if the esophagus was tied so as to prevent vomiting. His co-worker, Dr. Orfila,[2] held that large amounts of alum, at least five times that taken by Madam B., were daily administered to patients without any disturbance whatever, and certainly did not produce dilatation of the heart. He did not deny that she had experienced unpleasant symptoms from taking the alum, especially as she was suffering from an affection of the stomach. As a result of his testimony, the fine previously imposed on the apothecary who made the mistake was reduced by half.

The practice of the medical art in the first half of the nineteenth century, in relation to the administration of alum, was thus stated by Orfila in 1852:

"An adult man can swallow in a day, and without inconvenience, 4, 6, 8, and 10 grams of calcined alum dissolved in water. Boerhaave gave 4 grams of it at a time in intermittent fevers. Helvetius gave every four hours 2 grams of pills containing 1 gram 30 centigrams of calcined alum, making the dose of alum 7 grams 8 decigrams in twenty-four hours. M. Dumeril has often administered 4 grams of this salt in a day, dissolved in a diet-drink. Marc gave, in twenty-four hours, 500 grams of whey, in which was dissolved 8 grams of alum. Dr. Kapeler has given without inconvenience, in painter's colic, and in the epidemic disease known under the name of raphania, up to 24 grams of alum in the course of twenty-four hours, and sometimes he has administered 12 grams at once dissolved in 200 grams of vehicle: some of the individuals subjected to this medication were of a feeble constitution. The medicament has only very rarely induced nausea or vomiting, never epigastralgia; but it has frequently caused abundant stools."

ALUM IN FLOUR

The view that aluminum in food, and particularly alum, was injurious to health came into prominence in connection with its addition to flour to cover up inferiority. It is a fact that popular reaction to food adulteration may be secured more easily by a claim of unhealthfulness than by any other claim not excluding that of inferiority or that of unfair valuation by reason of such sophistication. Hence, the food reformer, be he government official, professional chemist, physician, or lay propagandist, in instituting a reform for correcting an abuse in food practices, finds it highly advantageous to emphasize the effects upon health. So great is the response to this appeal that there is always the temptation to draw hasty conclusions in the matter of wholesomeness and, from the viewpoint of the physiological facts, to pronounce the sophisticated food as actually injurious. There has been and can be little opportunity to correct such error, for to do so would be seemingly to champion fraud. When, however, the addition to food does not hide inferiority but serves a useful and justifiable purpose, the whole matter assumes a different aspect. It becomes important to learn the real truth, lest by erroneous condemnation the public lose the economic advantage of a wholly justifiable and useful product.

It appears that the claims of unwholesomeness in the case of the addition of alum to flour to conceal inferiority were not based upon any established knowledge of such unwhole-someness, yet passed unchallenged for the reasons just recited. Such additions for the purpose of fraud could not even seem to be championed, for it was a nefarious practice, condemned by the knowledge of its existence. Yet it is interesting that during the same time the addition of alum to pickles went unchallenged, first, because it was a domestic practice and second because it was serving a useful purpose. While, of course, the amount of alum that would be consumed in pickles was probably considerably less than would be consumed in flour, yet had the claims of injurious action put forward in connection with alum adulteration of flour actually been established, we might have expected a considerable reaction against its use in pickles. The reason why no such considerable reaction occurred is apparently that the usage in flour was wholly fraud, while in pickles the alum was serving a justifiable purpose.

Under these circumstances, we are unable, from the mere fact that claims of injurious effects on health grew out of the fraudulent addition of alum to flour, to get any real light upon the question as to whether such added alum was or was not actually harmful to health. Unfortunately, too, unbiased scientific research in this field was so scanty and so crude that it has passed on to us little knowledge of this subject. Further, clinical medicine gave no evidence of such injurious effects, although the addition of alum to flour was widely practiced.

That in other foods it might be injurious to health came to us almost wholly as an unsupported statement by those interested in correcting its fraudulent addition to flour. Thus, the English writer Hassall[3] in 1857 characterized it as particularly injurious, but did not mention any single case of injury or any specific way in which injury was produced. At the same period Taylor,[4] another English writer, presenting the subject from the toxicological viewpoint, which as a whole must be regarded as more impartial, wrote in reference to its ingestion in large quantities:

"We cannot, therefore, refuse to admit the fact of this substance acting as an irritant, on the principle on which we admit the irritant properties of salts of a far more innocent character. It is, however, proper to observe that alum given in large doses to animals does not appear to affect them seriously unless the gullet be tied; 3 drachms, dissolved in 6 ounces of liquid, have been taken at a dose without any inconvenience resulting."

This statement gives emphasis to the fact that a saline such as alum or even common table salt, when taken in quantity and in concentrated form, acts as an irritant and tends to induce vomiting; if vomiting is not produced, either because it is artificially prevented or for any other reason, the irritant action continues and may occasion serious consequences. It is in reference to this that Stillé,[5] stated in 1874:

"There is no case on record of death or even of material inconvenience from the internal use of this medicine [alum], in any dose, by a person previously free from serious disease. It is possible, however, that if from insensibility or any other cause the alum were not vomited, it might occasion grave lesions of the stomach.'

Whether or not he had considered Riquet's[6] case, reported a year previously, is not clear. In this instance 30 grams (not 30 grains as has been erroneously quoted) is reported to have been followed by death. It well may be that this substance was not alum, since other organs and not the stomach were analyzed and the description of the procedure is not convincing.

Another factor to be considered in connection with salines like alum is emphasized by Tardieu[7] when he states, "the [alum] concentration is of more importance than the dose." That is to say, a smaller amount in concentrated form may produce a certain irritating action while a greater amount in a dilute form may be without such irritating action. That this is true of common table salt is well known.

COMPOUNDS OF ALUMINUM IN BAKING POWDER

In the middle of the 19th century, long after alum had been condemned as an adulterant in flour, the baking powder idea was put into practice, namely the combining of baking soda and cream of tartar in exact proportions as a baking powder to replace the separate use of these ingredients in inexact proportions by the housewife. In the seventies, the search for some other "acid ingredient" than cream of tartar led to the temporary use of alum which later was replaced by s.a.s. This was used not only by competing manufacturers but also by the makers of cream of tartar powders who wished to produce cheaper brands to meet the competition of the "alum" powders of their competitors. At the same time, taking advantage of the situation that had developed as a result of the fraudulent use of alum as an adulterant of flour, claims of unhealthfulness of so-called "alum baking powders" were widely heralded in an effort to stem the tide of the commercial success that seems to have attended even the early sales of these powders, though the manufacturers of tartrate powders were themselves manufacturing and selling the powders which they denounced. Apparently as a result of such claims of unhealthfulness, the subject of the wholesomeness of these baking powders soon became a matter of scientific investigation.

RESEARCH

The work of G. E. Patrick,[8] professor of chemistry, University of Kansas, in 1879, is among the earliest of many physiological studies conducted in the past half century on food containing aluminum introduced as an ingredient of baking powder. So far as it appears, Patrick was a disinterested investigator who wished to determine at first hand the real facts in the case. His work consisted in the feeding of baking powder biscuits to cats, killing them at various stages of the digestive processes and examining the stomach and intestinal contents for dissolved aluminum. He failed to find aluminum in solution and concluded that the use of a properly compounded powder in making bread or biscuits was harmless to the human system.

In 1880, James West-Knights,[9] public analyst, published certain experimental observations on "The Action of Alum in Bread-Making" which presumably formed the basis of his opinions given in the Norfolk Baking Powder Case (p. 146). He stated:

"I prepared some gluten from pure wheat flour, and weighed four portions of 2 grams each, which were treated as follows:

"I was boiled ten minutes in pure distilled water.

"II was boiled ten minutes in a weak solution of alum.

"III was boiled ten minutes in weak alum and carbonate of soda solutions, with the resulting precipitate of hydrate of alumina.

"IV was boiled ten minutes in weak alum and phosphate of soda solutions, and the resulting precipitate of phosphate of alumina.

"Each sample, after washing with water, was digested in 50 cc. of an artificial gastric juice (consisting of pepsin and 0.2 per cent hydrochloric acid), at a temperature of from 90° to 96°F.; after five hours digestion, the residues that remained were removed, gently washed with distilled water, dried and weighed.

"I had entirely disappeared.

"II A tough spongy residue remained, which weighed 1.05 grams.

"III A similar residue remained, which weighed 0.90 grams.

"IV A similar residue, weighing 0.80 grams.

"These experiments have been repeated, using lactic acid instead of hydrochloric, with similar results.

"I think the inference can fairly be drawn from these results, that gluten, after treatment with alum, or insoluble salts of alumina, is less soluble than ordinary gluten, in the gastric juice, by about one-half. Whether the alumina is in a soluble or insoluble form seems to have no great influence on its effects upon the gluten.

"I next compared the digestibility of pure and alumed bread. The samples employed were:

"v Pure bread made with yeast.

"vi Bread made with same flour as above, and 'Alum Baking Powder' (containing about 30 per cent of alum), in the proportion of a teaspoonful of baking powder to a pound of flour.

"Two grams of crumb in each case were taken, in its natural state of moisture.

"After six hours digestion in 50 cc. of gastric juice at 90° to 96°F., the residues in each case were collected, washed, dried and weighed, with the following results:

"v Residue weighed 0.40 grams.

"vi Residue weighed 0.66 grams.

"If 40 be taken as the average percentage of moisture in bread, and that percentage deducted from the 2 grams originally taken, it leaves 1.2 grams of dry bread operated upon, of which, in the pure sample, 0.80 grams, corresponding to 66 per cent, was dissolved; and in the alumed bread 0.54 grams, or 45 per cent only.

"Or, in other words, the pure bread was one-third more soluble in the gastric juice than the bread containing the 'Alum Baking Powder.'

"I next tried the action of alum upon diastase.

"Two separate grams of crushed malt were weighed, and to one was added 0.1 grams of crystallized alum; both samples were then digested with 20 cc. of water at a temperature of 160°F., and maintained at that temperature half an hour, then filtered, and the residues washed with cold distilled water.

"The pure sample gave 0.70 grams of extract.

"The sample containing alum gave 0.185 grams only of extract, including the alum added, or that portion of it that was not retained in the reiidue. A considerable portion of the alum was in the extract, and no doubt a very much smaller quantity would have had the same effect.

"This experiment shows that alum exerts a very marked influence on the conversion of starch by diastase; as diastase is similar in its action to, and supposed to be identical with, ptyalin, the ferment of the saliva, I think this has a direct bearing upon the indigestibility of alumed bread; for not only is the gluten of the bread but also the starch rendered much more indigestible by the presence of alum.

"This powerful action of mere traces of alum or salts of alumina upon soluble gluten and diastase is, I think, sufficient foundation upon which to assert that alum, either in a soluble form or mixed with carbonate of soda, is injurious to health when introduced into bread; the extent of the injury may or may not be small."

The work of Henry A. Mott, Jr.,[10] a chemist of New York, published originally in the *Journal of the American Chemical Society* (1880) and republished some years later, appears to be the first elaborate effort to investigate the question of the healthfulness of food containing salts of aluminum introduced through baking powders. He cited the experiments of Devergie and Orfila in support of the toxicity of alum but failed to note that while the former reached the conclusion that alum was toxic, his co-worker Orfila concluded that alum was not poisonous as we ordinarily employ that term. He further cited the early case of Madam B., but did not state that the judgment was cut in half because of Orfila's testimony against the toxic action claimed. He cited the case of Dr. Ricquet, but in the original publication quoted the amount as 30 grains instead of 30 grams, very likely a typographical error. He put emphasis on the opinions of the injuriousness of alum added to flour as a fraudulent adulteration and on the fact that no scientists had come forward to recommend the use of alum in breadmaking. Finally, he approached the experimental part of his subject with the statement: "It hardly seems necessary for any experiments on animals to decide a question of this nature, so that the use of alum baking powders can be condemned; for a scientific consideration of the subject can lead to no other conclusion."

He experimented with dogs. In one series, they were fed solely on biscuits, usually eight at a time, made with 10 and 20 teaspoonfuls of alum baking powder to the quart of flour divided into twenty-seven and twenty-two biscuits. He reported loss of energy, weakness of the limbs, vomiting and constipation.

In another series, dogs were fed hydrate of alumina with meat, $\frac{1}{8}$ oz. to $\frac{3}{8}$ oz. per feeding. It was reported that vomiting occurred. Vomiting did not follow feeding with phosphate of aluminum.

In another series of observations digestion was conducted in vitro, making use of gastric juice obtained from dogs through gastric fistulae. Fibrin in the presence of ten times the quantity of moist alumina or of alum was only partially digested; white of egg, not at all.

A dog was fed moist hydrate of alumina in quantity and then this substance repeatedly introduced into the stomach through a tube over a period of some days. It was reported that alumina was found in the blood and liver. In another animal, fed 16 oz. of moist hydrate of aluminum, it was reported found in the blood, kidney, heart, liver and spleen in large quantities. The same results were obtained after feeding phosphate of aluminum. It was concluded that aluminum, instead of passing out of the system, accumulated in the various organs, interfering with their proper functions.

In 1886, were published, the results of experiments conducted by Paul Siem[11] under the direction of Hans Meyer. Throughout his experimental studies on animals he employed organic double salts of aluminum. This work is reported here, not because of any real bearing upon the subject of aluminum in food, but because it has been cited in justification of opinions that aluminum in food is injurious.

In his work on frogs, Siem avoided the double sodium salts with tartaric and citric acid, employed with mammals, since these salts caused disturbances with control animals, but used the double lactate $Al_2 \begin{cases} (C_3H_5O_3)_3 \\ (C_3H_4NaO_3)_3 \end{cases}$, containing 15.2 per cent Al_2O_3, injecting in solution into the dorsal lymph sac. Controls with sodium lactate, employing doses of 0.05 gm. to 0.1 gm. per medium-size frog, showed only local irritation for one to two minutes, while doses of 0.3 gm. to 0.4 gm. caused general paralysis with death in a few hours in one-half the animals, the other half recovering by the next day. Autopsy findings were not reported.

With sodium aluminum lactate (doses are in amounts of Al_2O_3 contained therein), 0.02 gm. to 0.03 gm. per medium-size frog usually caused death in ten to twenty-four hours, but in a few instances not until some days later. It will be noted that the amount of lactate actually administered was 0.13 gm. to 0.197 gm., or about one-half the dose that without aluminum produced death in one-half the control animals. Further, if it be assumed that a "medium-size frog" weighs 30 gm., the amount of aluminum oxide per kilogram body weight injected as a single dose was 7 gm. to 10 gm. in a soluble combination. Such a dose is equivalent to 1 lb. to 1½ lb. per 70 kg., the weight of a human adult, or 6500 to 9700 R. B. units.

In general, the following effects were noted: The frog was restless for a few minutes; motionless for four to six hours, but when touched jumped like a normal animal; there was gradual slowing of movements with feeble response to skin irritation, although induced current applied to the cord caused powerful tetanus of the posterior extremities. A few hours later, reflex excitability was lost, but the heart beat was powerful and rhythmic, finally stopping in diastole; it was not acted upon by atropine but contracted from mechanical irritation. He concluded that aluminum had a direct paralytic action on the ganglia of the brain, cord and heart. There were no reports of autopsies.

To rabbits, dogs and cats, the double tartrate of aluminum was administered, usually subcutaneously. The fatal dose (as Al_2O_3) per kilogram body weight was: Rabbits, 0.3 gm.; dogs, 0.25 gm.; cats, 0.25 gm. to 0.28 gm. given in ascending doses extending over two to four weeks. Symptoms generally appeared in from three to five days, and were referable to the gastrointestinal tract; they were loss of appetite and constipation. Later there was loss of weight, increasing emaciation and weakness, vomiting, psychic depression and loss of sensibility, low temperature, in many instances diarrhea and death, either without marked disturbances or preceded by respiratory disturbances and clonic convulsive seizures. During this time the urine was lessened but aluminum was not constantly present. The average lethal dose was lower if the initial dose was large; however, the intravenous administration of a single dose 4 or 5 times the above lethal doses failed to produce death in less than one to one and one-half weeks.

Post mortems: The mucous membrane of the stomach and small intestines showed slight hyperemia and swelling (though never to the extent of toxic gastroenteritis) with here and there small ulcers, especially in the gastric mucosa. The kidneys showed evidence of parenchymatous nephritis. The liver showed fatty degeneration. Siem considered the irritation of central origin, bulbar paralysis, since the muscles and nerves were normal.

The results of Siem's observations, even as demonstrations of the toxic action of the subcutaneous and intravenous administration of aluminum, are seriously impaired by the toxic character of the organic compounds with which the aluminum was combined. This is indicated by his own controls on frogs and has been demonstrated by later studies of the toxic action of tartrates.

In 1887, Lucius Pitkin, PH.D.,[12] chemist, New York City, reported results of investigations he had been called upon to make to determine the solubility in gastric juice of aluminum residues in baking powder biscuit.

He used a powder composed of sodium bicarbonate 29 parts, burnt ammonium alum 20 parts, calcium acid phosphate 20 parts and corn starch 65 parts, and used 13.615 gm. to a quart of flour, the resulting biscuit weighing 818 gm. Of this bread, 60 gm., estimated to contain 36.6 mg. Al_2O_3, was digested for three and one-half hours at $37\frac{1}{2}°$c. with 200 gm. canine (fistula) gastric juice; 23.2 mg. Al_2O_3 were recovered in the undissolved residue and 10.9 mg. in solution. Allowing for the difference in acid strength of the canine and human gastric juices, he calculated that the solubility of the aluminum in the mouth and stomach of man would be approximately one-fifth to one-sixth of the total amount present.

In 1889, Pitkin[13] reported the comparative inhibiting action on the digestive action of pepsin-hydrochloric acid (0.2 per cent) and egg albumen.

He added (1) 2 per cent and 1 per cent respectively Rochelle salts and (2) 0.2 per cent and 0.1 per cent respectively burnt ammonia alum using these relative quantities on the basis that the ratio of the soluble residues present in biscuit made with the corresponding powders was about 20 to 1. The inhibition by Rochelle salts was 60 per cent and 58 per cent respectively, and by the burnt alum 41 per cent and 24 per cent respectively.

It seems clear that for a fair comparison a larger quantity of soluble sulphate should have been present in the aluminum digestive mixture.

In 1887, R. F. Ruttan, B.A., M.D.,[14] published the results of experimental observations on the relative digestion in vitro of yeast (Y), tartrate (T), phosphate (P) and alum baking powder (A) breads.

With peptic digestion, the ratio of A to Y bread digested was 86.1 and 69.9 to 100; with tryptic digestion 58.7 and 54.3 to 100. The relative tryptic digestion of all varieties with variations in the time of exposure were:

TABLE XI

Time, hours	Y	T	P	A
1½	100	84.8	62.5	60.0
2½	100	93.7	88.6	88.7
3½	100	101.5	95.0	92.7

"Under the conditions of the experiment, the products of digestion accumulating in the beakers soon became a more powerful retarding agent than the constituents of the powders."

TABLE XII

AMYLOLYTIC ACTION (PER CENT OF STARCH CONVERTED)

	Y	T	P	A
Pancreatic	26.1	22.6	21.3	20.3
Salivary	27.0	19.3	17.8	17.0

Comparative tests with digestive mixtures to which were added 0.05 per cent of the respective powders yielded:

TABLE XIII

	Control	T	P	A
Saliva, 1 per cent starch, per cent converted..........................	26.6	24.0	19.2	18.8
Pepsin, HCl, gluten, relative digestion.	100	92.7	92.7	73.4
Pancreatic proteolysis, relative digestion...............................	100	90.0	89.1	83.4

In a series of experimental observations on the influence of salts produced by the decomposition of certain baking powders on the amylolytic action of saliva, expressing the result in per cent of starch converted, Ruttan reports for 1 per cent solutions: Rochelle salts 26, trisodium phosphate 24.6, sodium sulphate 24, potassium sulphate 22.4, aluminum hydrate 21.6, calcium sulphate 20.8 and acid sodium phosphate 19.6; for 0.05 per cent solutions: Rochelle salts 27.6, aluminum hydrate 24.4. Using different quantities of alum and other aluminum salts, the retarding action was in proportion to the amount of aluminum salts present. The retarding action of phosphate bread, Ruttan explained by the presence of the 13 per cent calcium sulphate in the powder used.

In 1888, there was published the result of experimental studies by Prof. J. W. Mallet,[15] University of Virginia, on the effects on digestion of the residues supposed to be left in bread from the use of alum baking powders.

He first investigated the qualities of seventeen brands of powders, finding that they contained a slight excess of sodium bicarbonate; two out of twenty-seven samples contained a slight excess of the acid constituent. He concluded that the aluminum in the residue from straight powders was in the form of the hydroxide; and from combination powders, nearly the whole was present as aluminum phosphate. Further, that the interior of the bread loaf baked in ovens with a temperature between 472° and 496°F. had a temperature ranging between 197° and 212°F., at which latter temperature 3 or 4 molecules of aluminum hydroxide retained 1 molecule of surplus water and 1 molecule of normal aluminum phosphate retained a little over 3 molecules of water.

His studies on digestion were made with these supposed residual compounds dried at 212°F. A weighed quantity (10 to 50 gr. of the hydroxide and 10 to 100 gr. of the phosphate) was swallowed ten to fifteen minutes before a meal, using only a little water, or along with the meal, and the effects noted, the amounts ingested being greater than would be derived from the quantity of bread usually eaten. On two or three occasions, particularly with the smallest amounts, there was no clearly observable effect; there was never gastric pain, but particularly with the larger amounts, "the well-known oppressive sensations of indigestion properly so-called, lasting for a longer or shorter time, but generally at least two or three hours after the taking of food."

A few experiments were made with weak hydrochloric acid-pepsin solutions, from which it was concluded that, when aluminum hydroxide, as above, was added, from 47.3 to 61.8 per cent was dissolved and at the same time from 5.6 to 14.5 per cent of the pepsin precipitated from solution; and that when aluminum phosphate, as above, was added, from 30.2 to 49.1 per cent was dissolved and 25.8 to 32.9 per cent of the pepsin rendered insoluble.

Considering all the results, the author concluded that the residues left in bread by the baking powders could not be viewed as harmless, but must be ranked as objectionable.

In 1892, Otto Hehner,[16] President of the Society of Public Analysts (England), published an account of experiments which he had conducted "to bring a case into court, in the hope of obtaining a reversal of the Cambridge (Norfolk Baking Powder Case) decision."

The powder used, calculated from an analysis, contained: Crystallized alum, 45.80 per cent, bicarbonate of soda 18.71 per cent; starch 33.40 per cent; and moisture and not determined 2.08 per cent. Theoretically, when the powder was mixed with water, the sodium bicarbonate should precipitate 35.19 parts of alum, leaving a considerable amount in solution. While a marked trace was left in solution, its quantity was less than required by theory.

The bread used was made with 11.21 gm. of powder to a pound of flour.

Experiments on the digestion of egg albumen were conducted in a fluid containing 2 gm. coagulated egg white, 1 mg. pepsin and

0.2 per cent HCl kept at 50°c. Insoluble nitrogen in the residue was determined by the Kjeldahl process. Blanks showed complete digestion. In the following experiments the indicated quantities of baking powder or alum were added to the digestion mixture.

TABLE XIV

Experiment	Baking powder, gm.	Time, hrs.	Per cent insoluble
1	.05	5	14.0
3	.05	5	28.0
5	.20	6	46.7
	Crystallized alum gm.		
2	.025	5	13.8
4	.025	5	25.6
6	.100	6	39.2
7	.100	6	52.7

It was concluded that "the same quantities of alum, either taken pure or in the form of baking powder, prevent to an equal extent the digestion of hard-boiled white of egg."

Similar experiments were conducted on the digestion of wheat flour. Blank showed 12.9 per cent of the albuminoids undissolved.

TABLE XV

Experiment	Baking powder, gm.	Time, hrs.	Per cent insoluble
8	.05	6	14.9
9	.20	6	14.9
	Crystallized alum gm.		
10	.025	6	34.4
11	.100	6	54.9

"In the case of digestion of flour, therefore, alumed baking powder has far less influence than the corresponding amount of alum contained in it. Alum itself has a most injurious influence upon the digestion of flour, while that of alumed baking powder is slight."

Experiments on the digestion of the crumb of bread, assumed to contain 7 per cent albuminoids, with which the indicated amounts of baking powder and alum were mixed, showed:

TABLE XVI

Experiment	Baking powder, gm.	Time, hrs.	Per cent insoluble
12	.....	6	16.5
13	.05	..	22.5
14	.20	..	29.1
	Crystallized alum gm.		
15	.025	..	22.5
16	.100	..	36.9

"Curiously, therefore, the influence of baking powder containing alum upon the digestion of bread is more marked than in the case of the raw flour. With small amounts, or rather the amount of baking powder recommended to be taken by the manufacturer, the influence of alum and of alumed baking powder is equal, but with larger quantities the alum acts more detrimentally than does the baking powder."

Digestion of 5 cc. milk, assumed to contain 3.66 per cent albuminoids, for six hours, showed:

TABLE XVII

Experiments	Baking powder, gm.	Per cent insoluble
15	.....	63.5
16	.05	93
17	.20	93
	Crystallized alum gm.	
18	.025	76.7
19	.100	76.7

"It is remarkable that in the digestion of milk the alumed baking powder exerts a more injurious influence than does the amount of alum contained in it, and the smaller amount of alum, whether taken as baking powder or as crystallized alum, acts as markedly as does the four times larger quantity."

From Hehner's data on the digestion of milk the author calculates the per cent insoluble to be 25.4, 37.2, 37.3, 30.7 and 30.7 respectively instead of the above amounts given by Hehner.

To each of 4 individuals 2 gm. of the baking powder, "very nearly the quantity in 4 oz. of bread," mixed with water and sugar were given. They each reported unpleasant symptoms beginning in half an hour, described by one as great weight in the region of the stomach, later epigastric pains, slight difficulty in breathing, headache, and ultimately slight diarrhea with discomfort for several days. Similar symptoms, almost equal in degree, were reported by the same individuals for 1 gm. of the powder after complete recovery from the first administration.

Hehner concluded that the powder exerted a most injurious influence on digestion and constituted adulteration injurious to health.

In 1894, there were published the results of studies on "The Influence of Alum, Aluminum Hydroxide and Aluminum Phosphate, on the Digestibility of Bread," by W. D. Bigelow and C. C. Hamilton.[17]

The dried bread used was yeast bread, sliced, dried at 98°, ground and bottled.

Methods: 1. Digestion in pepsin solution. Pepsin (Merck's granulated) 0.1 gm.; 0.33 per cent HCl 100 cc.; dried bread 2 gm.; temperature 40°c.; time, twelve hours. Residue kjeldahled.

2. Stutzer's method: 2 gm. portions of dried bread were digested in artificial pepsin solution as above; the washed residue digested six hours in 100 cc. Stutzer pancreas solution; the residue kjeldahled.

3. Stutzer's method modified by Wilson (digestion with pepsin as above followed by the use of artificial pancreatic solution).

4. Neibling's method: Acid (0.2 per cent HCl) hydrolysis followed by pancreatic digestion, using Stutzer's pancreatic solution.

5. Neibling's method, modified (same as preceeding, excepting the use of the artificial pancreatic solution).

The results obtained are shown in the following table:

TABLE XVIII

PER CENT ALBUMINOIDS DIGESTED

Method	1	2	3	4	5
Blank..........................	93.26	93.57	93.21	93.28	93.21
Alum added to flour, per 2 lb. loaf 0.8 gm........................	89.11	92.56	92.21	92.54	92.74
Alum added to flour, per 2 lb. loaf 4.28 gm.......................	80.98	92.40	92.44	.....	92.02
Aluminum hydroxide added to flour per 2 lb. loaf 0.54 gm.......	87.03	92.18	92.00	91.77	91.90
Aluminum hydroxide added to flour per 2 lb. loaf 2.50 gm.......	86.78	90.43	90.21	89.13	88.96
Aluminum phosphate added to flour per 2 lb. loaf 0.64 gm.......	80.87	82.56	86.35	86.35	86.46
Aluminum phosphate added to flour per 2 lb. loaf 3.20 gm......	71.21	81.32	81.32	82.18	81.74

The bread was the yeast bread described except the additions to the flour noted.

Conclusions: The custom of judging the influence of alum on the digestion by its effect on the digestion in the pepsin solution alone has led to grave errors. The presence of alum interferes materially with the gastric digestion; a portion of the digestion which should be affected by the pepsin is affected by the pancreatic fluid.

With aluminum hydroxide, likewise, the albuminoids whose digestion is prevented in the pepsin solution by the dissolved aluminum compounds seem to be almost all digested by the alkaline pancreas solution. The action of aluminum phosphate is quite different, however, for notwithstanding the supposed insolubility of this compound, 10 to 12 per cent of the albuminoids which are digestible in the presence of alum or aluminum hydroxide appear to be insoluble in the presence of an equivalent amount of the phosphate.

In 1897, Döllken[18] repeated and extended the work of Siem on the toxic action of sodium aluminum tartrate, administered subcutaneously to animals. Poisoning by large doses, administered consecutively at short intervals, led to the same results, except that his cats and dogs showed central stimulation a longer time before death, for which reason he was unable to account for it on the ground of respiratory disturbance.

Variations in the aluminum administration permitted many details of the symptom complex to be brought out more acutely and changed the toxic manifestation in certain respects. He used a neutral solution containing 0.02 gm. Al_2O_3 per cubic centimeter.

Rabbits given 0.015 gm. NaAl tartrate (probably as Al_2O_3) lived twelve to twenty-five days. Other rabbits given 0.005 NaAl tartrate subcutaneously per day, showed loss of weight in six to eight days; death in fourteen to thirty-five days. Guinea pigs were very resistant. In two hundred and fifty days they had received Al_2O_3 0.12 gm. and flourished; they were given 0.007 gm. every five days; in thirty days they showed loss of weight and died in fifty days. Dogs and guinea pigs received in one year ten times the fatal dose without symptoms. Results in this respect differed from those in lead poisoning. There was a great difference in the reactions of animals of the same species. Dosage in cats of 1 mg. per kg. every two to three days subcutaneously for six weeks, resulted in no symptoms; two to five mg. per kg. every two days produced death in forty days, loss of weight, bulbar symptoms, peculiar gait. If aluminum was then stopped, they recovered. There were depression, stiff extremities, swaying, difficult tongue movement, hoarseness, loss of appetite, lowering of psychical functions; voluntary movements became impossible, the animal lay on its side, sensibility was lowered, clonic and tetanic convulsions occurred, especially in the anterior half of the body.

Dogs acted in much the same way. One received a total of 1.8 gm. Al_2O_3 subcutaneously over a period of about a year with a seven weeks' intermission. Later with larger doses he showed much the same symptoms as those of the cat. Relatively large doses caused loss of weight probably due to the toxic enteritis.

The pathology of the central nervous system was studied in detail. Gross lesions were limited to cloudiness of the pia and venous congestion of the brain and cord.

Microscopically, the white matter of the brain showed no pathological change, and that of the cord relatively little, limited to slight degenerative changes, most marked in the posterior tracts of the lumbar region, less in the posterior and lateral tracts of the cervical cord and only very slight in the dorsal tracts.

The motor root fibers of the fifth, seventh and twelfth cranial nerves showed degeneration, as did also fibers in the ninth, tenth and eleventh.

Nerve cells showed various stages of degeneration, viz.,

1. Hyaline, with cloudy appearance, diffuse staining, with many normal cells.

2. Fine granular, with many cells undersized; nuclei might be lacking; granules gathered at nuclei or cell margin stained with difficulty.

3. A further development of No. 2 with a crescent of granules or fibrils at the cell margin, the majority of cells without nuclei.

The cells of the cord showing pathological changes were most numerous in the lumbar, less in the cervical and relatively few in the dorsal cord, the cells of the anterior cornua being most markedly affected, and those in the interior slightly, the stages of degeneration varying in the particular animal and with the duration and severity of the toxic action. The cells constituting the nuclei of origin of the cranial nerves, five to twelve, showed similar degenerations of varying intensity.

Capillary hemorrhages were observed in the gray matter of the cord.

V

RESEARCHES BY E. E. SMITH, 1898–1902

During the period between 1898 and 1902, E. E. Smith conducted a series of investigations at the request of the Missouri, later the American, Baking Powder Association on the broad question of healthfulness of food prepared with so-called alum baking powder. Throughout the work Austin Flint, late professor of physiology (at Bellevue Hospital Medical College and at Cornell University Medical College), and author of works on physiology, and Peter T. Austen, PH.D., formerly professor of chemistry, Rutgers College, cooperated in an advisory capacity. Since they accepted in full his original plans both for the scope of the investigations and for the methods of procedure, the responsibility for the work rested independently with the writer.

In arranging for this work, the representatives of the Association gave emphasis to the necessity of frankly and fully establishing the facts in the case, since, on the one hand, if such food were in truth in any way unwholesome, it was regarded as essential to good business policy that the baking powder producers at once modify the composition of their powders; while, on the other hand, if such food were not unwholesome, that they defend their products, in refutation of the claims of unwholesomeness advanced by their business competitors.

In considering medical and other scientific evidence and opinions expressed in the twenty-five years, more or less, during which the matter had been in controversy, it was observed that claims of unwholesomeness were entirely inferential. No case of injury from the consumption of such food had ever been recognized, and rarely, if ever, had instances of supposed injury been reported, excepting in "blind" newspaper advertising. Further, except in its empirical use, the results of which were entirely negative, no direct studies of the effects of such food had been made.

It was interesting that in the early years of the controversy, theoretical objections were based upon the view that aluminum deprived the body of phosphate and rendered gluten and dextrin indigestible. Subsequently the effects on phosphates were ignored and it was held that aluminum, instead of being rendered insoluble by the phosphates and thus passing out of the system, accumulated in the various organs, interfering with their proper function, and, also, that it interfered with digestion through inhibition of the activity of the digestive enzymes.

Accordingly, the writer planned his investigations to determine by direct and exact observation the extent to which food prepared with the so-called alum baking powders was utilized by the body and whether or not there were accumulations of aluminum and interference with the digestive and metabolic functions.

In a study of the literature of the subject no record was found of exact physiological observations upon the use of such food. Opinions regarding its healthfulness were usually based upon what was believed regarding the conditions of the alumina into which the "alum" of the powder had been converted in the preparation of the food. According as it was believed to be soluble or insoluble, the food was thought to be harmful or innocuous. The alleged harmfulness was in the production of disturbances of the digestive functions, resulting from the action of soluble aluminum salts, and in the diminished digestibility of the food product. In view of the uncertainty of the chemical conditions and of the consequent diversity of opinion, it seemed to the writer that direct physiological observations of the use of the food would be of very great value and must be given the final consideration in reaching a conclusion as to its wholesomeness. The following experiments were accordingly planned and executed and were the evidence upon which Austin Flint and the writer based their testimony as witnesses for the defendant in the trial in the case State of Missouri vs. Layton.[1]

"The food the use of which was the subject of these observations was bread prepared according to the following recipe:

BAKING POWDER BREAD (RECIPE NO. 1)

Flour	585	gm.
Salt	8	gm.
Alum baking powder	17¼	gm.
	(2 teaspoonfuls)	
Water	500	gm.

"Directions: Sift the dry ingredients together three times, stir in the water, and bake in one loaf at 175° to 185°c. for an hour and three-quarters. (The baking powder used was of average quality, made in the regular process of manufacture, being substantially a combination of the following: sodium bicarbonate, one part; desiccated sodium aluminum sulphate, one part; starch, three parts.)

"Bread for use in control experiments was prepared without the use of baking powder in a manner as nearly as possible like the preceding and so as to yield a product in other respects chemically and physiologically identical with it. This was accomplished by the use of the method of Liebig, after the following recipe:

CONTROL BREAD (RECIPE NO. 2)

Flour	585	gm.
Salt	8	gm.
Sodium bicarbonate	3½	gm.
Water	489	gm.
Hydrochloric acid (15.5 per cent)	11	cc.

"Directions: Sift the ingredients together three times, stir in the water and acid, and bake in one loaf at 175° to 185°c. for an hour and three-quarters. (The amounts of sodium bicarbonate and acid were so regulated as to yield the same amount of gas for raising the bread as was yielded by the powder in formula No. 1.)

"The first set of experiments was planned to determine the influence of this food on secretion in the stomach. The experiments consisted in the administration of test breakfasts of baking powder bread and control bread to healthy individuals and the subsequent withdrawal of the stomach contents after one hour, precisely in the manner familiar to the profession from the use of the method for the purpose of determining the secretory power of the stomach in suspected conditions of disease of this organ. The two individuals used as the subjects of these experiments were healthy young men. After eating dinner the night previous to the test, they partook of no food till the test breakfast was administered the following morning. The test meal was the same with each subject and consisted of bread, 60 gm., and water, 350 cc., the simple Ewald breakfast. The stomach contents were withdrawn in exactly one hour, in the usual way, with a tube and without the addition of water. Four days later, a second test breakfast was administered in each case under precisely similar conditions, excepting the bread used. The stomach contents were examined for acid and pepsin, with the following results:*

TABLE XIX

SUBJECT I

	Baking powder bread	Control bread
Total acidity, as HCl......................	0.251 per cent	0.233 per cent
Total hydrochloric acid....................	0.215 per cent	0.222 per cent
Combined hydrochloric acid...............	0.143 per cent	0.136 per cent
Free hydrochloric acid....................	0.072 per cent	0.086 per cent
Pepsin test..............................	1 hour	1½ hours

SUBJECT R

	Baking powder bread	Control bread
Total acidity, as HCl......................	0.220 per cent	0.221 per cent
Total hydrochloric acid....................	0.215 per cent	0.220 per cent
Combined hydrochloric acid...............	0.145 per cent	0.131 per cent
Free hydrochloric acid....................	0.070 per cent	0.069 per cent
Pepsin test..............................	1½ hours	1½ hours

"It is the writer's opinion that these results are negative; that the slight differences are the inevitable variations in physio-

* "The methods used were as follows: Total acidity, titration to a deep color with standard potassium hydroxide, phenolphthaleine indicator; total hydrochloric acid, method of Hayem and Winter; free hydrochloric acid, titration to a pure yellow with standard potassium hydroxide, dimethylamidoazobenzol indicator; combined hydrochloric acid, by difference; pepsin test, 0.2 gm. of moist egg-albumen, made by boiling eggs fifteen minutes and passing the whites through a 40-mesh sieve, was mixed with 5 cc. of filtered stomach contents, the whole kept at 38° to 40°c., and the time noted when the albumen was completely dissolved."

logical experiments observed even when the physical conditions appear to be identical. Such parallel experiments, separated necessarily by time, are doubtless subject to slight variations owing to different physiological states of the subjects impossible to control, but which are without material effect. The results obtained are well within these limits of physiological variation.

"There are some objections that have been offered to the foregoing experiments. It is stated that the amount of the test meal was too little. Sixty gm. were used because it is the amount that has been found most advantageous for stimulating the gastric secretion for purposes of diagnosis. It is about the amount of bread that many individuals take with a meal, and it is also a sufficient quantity to stimulate well the gastric secretion.

"The criticism is offered that the amount of gastric juice secreted was not measured. It was, of course, impossible to do, and the omission is not fatal to the interpretation of the results, for when it is remembered that we know the degree of acidity to which the more than 350 cc. of stomach contents were brought in each instance by the addition of the secreted juice, it must be concluded that in this we have an indirect measure of the quantity of secretion that is quite sufficient to enable us to judge how the quantities compare in the two parallel experiments. The statement that the employment of the stomach-tube stimulates a sudden gush of gastric juice and is responsible for the secretion calls for serious consideration. The process of secretion is slow even under very strong stimulation. There is no reserve of secreted fluid that can be suddenly set free, and the above-mentioned accident in the experiments is quite inconceivable.

"Attention has been directed to the fact that the pepsin test was not a quantitative estimation of the amount of pepsin present. The limited amount of material did not permit of a quantitative estimation of the pepsin in the stomach contents, so that our knowledge of this constituent is limited to the fact of its presence and to the fact that it exhibited the same activity in a simple test, the chief value of which is qualitative, but in which we might reasonably hope for some indication if the proteolytic activity was materially different."

The digestibility of the nitrogenous matter in the baking powder and control breads was determined by the Sjollema modification of the Stutzer method as follows: The crumb was dried at 38° to 40°c. for twenty-seven hours, finely divided in a mortar and the particles used for the test which passed through a 10-mesh but not through a 20-mesh sieve. The following mixtures were prepared:

Bread 2 gm.

Distilled water 380 cc.

2 per cent pepsin solution (Fairchild). 50 cc.

Hydrochloric acid sp. gr. 1.048. 16 cc.

These mixtures together with a control were uniformly incubated in flasks at 38° to 40°c. with frequent agitating and at the ends of sixteen, twenty-four and forty hours 11 cc. of the above hydrochloric acid added to each. At the end of forty-eight hours the contents of each flask were made up to 500 cc. by the addition of distilled water and the amounts of nitrogen in solution and undissolved determined. It was found that 99.0 per cent of the nitrogenous constituents of the control bread and 99.2 per cent of the nitrogenous constituents of the baking powder bread had been dissolved.

"A more elaborate experiment was then conducted to determine whether the baking powder bread was absorbed from the alimentary tract to the same extent as the control bread, and whether in the process of digestion and absorption there was evidence of any disturbing influence. The procedure was based upon the method for determining the coefficient of availability of the constituents of a dietary. An individual was given a definite diet for a certain period of time, in this experiment three days, and from the amount and composition of the food eaten and the stools separated it was determined how much the body had gained from the diet. By comparison with a control period under precisely similar conditions, only substituting control bread, it was possible to conclude from our experiments whether or not the baking powder bread was utilized by the body to the same extent as the control bread. The inquiry further included a study of the urine collected during each period, to determine the amount of nitrogenous waste, and the relative amount of the products of putrefaction absorbed from the intestine. The subject was a healthy man. At times he had experienced sensations of distress in the gastric region, but always following indiscretions in diet, such as eating sweets in large quantities. Under ordinary conditions he did not experience symptoms of stomach dyspepsia.

"The diet during the experiment consisted exclusively of bread, meat, milk, and butter as described. The bread used was the crumb of the baking powder bread and control bread previously described. This was chopped and weighed off in suitable portions for an entire period and sampled for analysis. The meat was lean beef selected from the round, chopped, cooked by heating in a frying pan, salt only being added, squeezed slightly in a press to remove a portion of the juice, and finally distributed into jars and heated in a sterilizer twenty minutes for each of three successive days. Several jars were used for analysis. The milk was obtained fresh and sampled daily, the composite sample for each period being analyzed. Butter was weighed off as required for use from a pail of butter that had been sampled for analysis. The analysis, made by the usual methods, showed the following composition of the food:

TABLE XX

	Total solids, per cent	Fat, per cent	Proteid, per cent	Carbohydrates, per cent
Bread, baking powder...........	47.54	0.22	6.50	39.40
Bread, control.................	49.74	0.24	6.81	41.30
Meat..........................	49.46	5.43	41.56	.....
Milk (baking powder period).....	12.34	2.70	3.10	4.78
Milk (control period)............	13.63	5.03	3.13	4.76
Butter.........................	89.55	88.04	0.49	.....

"Period 1. The baking powder period was of three days' duration. At the end of the night meal of the day before the period, milk was drunk and 6 grains of lampblack were taken in capsules, and thereafter till the morning of the fourth day following nothing was eaten excepting the prescribed diet, and only water was drunk. The morning meal of the fourth day, the first following the end of the period, consisted of a pint of milk, which was preceded by the administration of 12 grains of lampblack.

"Period 2. The control period was begun just one week from the beginning of Period 1, and was conducted exactly as the previous period, with the exception of differences noted in the dietary, which is given below (in grams).

DIET

Period 1.

    Baking powder bread.................................. 1800
    Meat.. 300
    Milk.. 1500
    Butter.. 150

Period 2.

    Control bread....................................... 1800
    Meat.. 300
    Milk.. 1500
    Butter.. 150

"The nutrients in the diet of each period were, in grams:

TABLE XXI

PERIOD 1 (BAKING POWDER)

	Solids	Protein	Fat	Carbo-hydrates
Bread..................	855.7	117.0	3.96	709.2
Meat...................	148.38	124.68	16.29	.....
Milk...................	185.1	46.5	55.5	71.7
Butter.................	134.325	0.735	132.06	.....
Totals.................	1323.505	288.915	207.81	780.9

"The baking-powder bread of this period made up 64.6 per cent of the solids eaten.

TABLE XXII

PERIOD 2 (CONTROL)

	Solids	Protein	Fat	Carbo-hydrates
Bread..................	895.32	122.58	4.32	743.4
Meat...................	148.38	124.68	16.29	
Milk...................	204.45	46.95	75.45	71.4
Butter.................	134.325	0.735	132.06	
Totals.................	1382.475	294.945	228.12	814.8

"The control bread of this period made up 64.7 per cent of the solids eaten.

"During each period the feces and urine were collected. The separation of the feces was satisfactory; they were slightly acidified with hydrochloric acid before drying.

TABLE XXIII
COMPOSITION OF FECES

	Period 1		Period 2	
	Percentage	Total grams	Percentage	Total grams
Air-dry material.............	.....	59.1	.....	68.3
Solids......................	90.38	54.006	83.08	56.744
Protein.....................	44.62	26.37	43.63	29.799
Fat........................	18.91	11.176	18.69	12.77
Carbohydrates..............	25.83	15.265	19.40	13.25

TABLE XXIV
COMPOSITION OF URINE

	Period 1	Period 2
Volume........................	3500 cc.	3250 cc.
Specific gravity....................	1.023	1.024
Nitrogen.........................	47.18 gm.	46.67 gm.
Indican..........................	Moderate amount	Moderate amount
Combined sulphate (H_2SO_4)........	0.556 gm.	0.528 gm.

"Comparison of the food and the feces in the two periods, expressed as percentage of the total diet:

TABLE XXV

	Solids	Protein	Fat	Carbohydrates
Feces, Period 1.................	4.08	9.1	5.37	1.95
Feces, Period 2.................	4.1	10.1	5.6	1.6
Diet available, Period 1..........	95.92	90.9	94.63	98.05
Diet available, Period 2..........	95.9	89.9	94.4	98.4

"It was thus found that the availabilities* of the diets in the two periods were practically identical, agreeing as closely, in fact, as duplicate experiments on precisely the same diet.

* "The stools consist of (1) metabolic products derived from the alimentary tract and its secretions, and (2) food residue from the diet. The diet less (2) is the digestible portion; the diet less (1) and (2) is the available portion."

"The compositions of the urine in the two periods, as regards the constituents indicating the degree of absorption of such aromatic products of putrefaction as are formed in the intestines, were so nearly the same that they did not indicate a greater degree of the putrefactive process in one or the other period. In fact, the evidence of the experiment was that the dietaries in Periods 1 and 2 were physiologically identical, and, since the dietary in each instance was composed to the extent of two-thirds of the bread described, the evidence of the experiment was that the two breads were physiologically the same.

"There are certain criticisms of this experiment well worth considering. The point is raised that metabolic nitrogen is not subtracted from the total nitrogen of the feces; in other words, that the figures obtained represent, not the actual digestibility, but rather the availability of the diet. It would be interesting to subtract the metabolic nitrogen and obtain actual figures of digestibility, but that is not usually attempted in such experiments, since the methods of determining the metabolic nitrogen at present known are quite inaccurate. It is, after all, the gain to the body in each period which we seek to establish, and that is, of course, indicated by the availability, and not by the digestibility.

"The question whether the diet was full and nutritious, including that of whether the subject was in nitrogenous equilibrium, is not to be given the same consideration here as in metabolism experiments in which the question of the balance of income and outgo is all-important. Of course, the diet should contain a fair amount of food, but no effort need be made to establish an equilibrium; on the contrary, to diminish the monotony of a fixed diet, investigators usually administer a little less than sufficient to establish an equilibrium. This does not put the subject in an unhealthy or in any way abnormal state. In fact, individuals vary their diet in this way in everyday life, and even some athletes, in training, for a time do precisely this thing to attain perfect health. The body adapts itself to certain variations that we must recognize as physiological, and the fact that today we eat less and tomorrow more is not evidence of ill health.

"In conclusion, then, the evidence of these experiments is that food prepared by the use of a so-called alum baking powder does not interfere with secretion in the stomach; and, even when it makes up the major part of the diet, it is utilized by the body in the same way and to the same extent as an acceptable control diet. The investigation does not reveal any reason for believing such food at all injurious or unwholesome."

Subsequent to the Missouri case, further studies were made:

FEEDING EXPERIMENTS WITH WHITE RATS[2]

On October 19, 1900, six white rats, four weeks old and from the same litter, were obtained and divided into two groups, I and II, which were kept in separate cages. Group I was fed liberally on baking powder bread, at first in milk, later moistened only with milk and finally with water, the bread being made by the use of two full teaspoonfuls of "Parrot and Monkey baking powder" to one quart of flour.

Group II was fed exactly as was Group I, excepting that the bread used was made after the Liebig formula.

At the beginning of these observations each group of animals was weighed and each week during the experiment their weight and conditions were recorded. The following results were obtained:

TABLE XXVI

Date	Group I baking powder, gm.	Group II control, gm.
Oct. 19....................................	55	56
Oct. 26....................................	61	62
Nov. 2....................................	90	95
Nov. 9....................................	143	146
Nov. 16....................................	156	177
Nov. 23....................................	208	222
Nov. 30....................................	228	254
Dec. 7....................................	276	293
Dec. 15....................................	303	324
Dec. 21....................................	324	345
Dec. 28....................................	334	361
Jan. 4....................................	343	350
Jan. 11....................................	338	353
Jan. 18....................................	323	358

One of the rats of Group II (control) was found dead on the morning of January 18. Its weight was 130 gm. Autopsy did not reveal the cause of death. Group II contained one large animal to which the difference in weight of the two groups seemed to be due.

FEEDING EXPERIMENTS WITH DOMESTIC PIGS[2]

On October 4, 1900, 4 boar pigs from the same litter about seven weeks old were obtained, which had never been taken from the mother, and had never been fed on swill. Their weights were as follows:

No. 1	45 lb.
No. 2	25½ lb.
Total	70½ lb.
No. 3	34 lb.
No. 4	39 lb.
Total	73 lb.

These pigs were kept in two separate pens. Nos. 1 and 2, known as control pigs, were kept in one pen, and Nos. 3 and 4, known as baking powder pigs, in another. During the continuation of these experiments the pigs were fed on different diets, the control pigs receiving a denfiite amount of bread made after the Liebig formula, together with a definite amount of milk, while the baking powder pigs received the same amount of bread made with a straight so-called alum baking powder, together with the same amount of milk as was fed to the control pigs. In making the bread the same ingredients in exactly the same amounts were used, excepting the material for raising.

The cereals of which the bread was made, consisted in each instance of equal amounts of hominy and flour. This made a coarse bread suitable for continuous feeding to animals. Great care was observed that even traces of baking powder bread should not become mixed with the control bread. To prevent this, separate dishes, spoons, and other kitchen utensils were employed for mixing and baking the breads, and the finished products were kept in separate boxes. Likewise for each set of pigs, separate pails were used in which the bread and milk were mixed, and caution was even taken to employ separate brooms for cleaning out the troughs.

In this way it was rendered absolutely certain that the control pigs did not receive baking powder or its products.

From October 4 to December 11, when the weights of the pigs were again taken, the animals received a total of 444 loaves of bread and 890 quarts of milk, being an average of 111 loaves of bread and 222½ quarts of milk to each individual pig; that is, for each day an average of about 1⅔ loaves of bread for each animal, and 3⅓ quarts of milk. Of course the actual amounts were gradually increasing quantities, varying with the age and development of the animals, each one receiving at the outset, when quite small, ½ a loaf of bread and 2 quarts of milk. These amounts were increased, so that on December 11 each animal was receiving 2 loaves of bread and 3½ quarts of milk. From these figures it is seen that the increase in the amount of bread fed was much greater than the increase in the amount of milk, the former being quadrupled, while the latter was not doubled. This was owing, on the one hand, to the desire to feed large quantities of bread, and on the other hand, to the increasing ability of the growing animals to consume large quantities of the solid food.

After December 11 there was a further increase in the amounts of the food fed, and on December 26 each animal was receiving on an average 2½ loaves of bread and 3½ quarts of milk. Up to this time the total quantities of bread and milk fed to the 4 animals were 588 loaves of bread, and 1048 quarts of milk, an average total for each animal of 147 loaves of bread and 264 quarts of milk. The recipe for the control bread was:

Hominy...	500 cc.
Flour..	500 cc.
Salt...	2 gm.
Sodium bicarbonate.............................	5 gm.
Dilute hydrochloric acid	
Water	

The hydrochloric acid was a dilute solution of just such strength and amount as was necessary to react completely with 5 gm. of sodium bicarbonate, and leave the solution just slightly acid to cochineal. The dry ingredients were sifted together three times, the acid and water together stirred in, and the loaves immediately placed in the oven to bake.

The baking powder bread was made after the following recipe:

Hominy...........................	500 cc.
Flour.............................	500 cc.
Salt..............................	2 gm.
Baking powder....................	(2 heaping teaspoonfuls)
Water	

The directions for use of the baking powder called for one heaping teaspoonful for a quart of flour. It will be seen that in the bread used in these experiments double the quantity of baking powder was used. It was calculated that each spoonful amounted to 15 gm. of baking powder, so that each loaf was made with 30 gm. of powder. Since animals Nos. 3 and 4 each ate an average total of 147 loaves of this bread during the period from October 5 to December 26, the bread eaten by each animal was made with a total of 4410 gm. of powder, about 10 lb. Analysis of the baking powder yielded 6.87 per cent of aluminum figured as Al_2O_3. Therefore the 4410 gm. of powder in the bread fed to each pig contained a total of 303 gm. (about ⅔ lb.) of aluminum, figured as Al_2O_3.

It is recognized that the animals were receiving an enormously greater amount of aluminic oxide under these conditions than would ordinarily be consumed, but in arranging the diet this was purposely made so, in order that the test might be rigid and yet be a purely physiological procedure and a test, not of aluminum compounds as such, but of food prepared by the use of a particular type of baking powder, which was the subject of inquiry undertaken.

On Dec. 11 the weights of the pigs were:

Control Pigs	Weight, lb.	Increase Per Cent
No. 1...	134.5	198.9
No. 2...	95.0	272.5
Total..	229.5	

Baking Powder Pigs	Weight, lb.	Increase Per Cent
No. 3...	108.5	219.1
No. 4...	125.0	220.5
Total..	233.5	

On December 26 the weights were:

Control Pigs	Weight, Lb.	Increase Per Cent
No. 1..	151.5	236.6
No. 2..	107.0	319.6
Total..	258.5	

Baking Powder Pigs	Weight, Lb.	Increase Per Cent
No. 3..	116.5	242.6
No. 4..	141.5	262.8
Total..	258.0	

During the continuation of these experiments the healthfulness and vigor of the control and baking powder pigs were normal, in each case being up to the standard for these animals. There was the usual voraciousness, without the pigs refusing at any time the rapidly increased supply of food. There was nothing to indicate that one food was more or less healthful than was the other.

On December 26 Pigs Nos. 1 and 4 were killed. An inspection of the internal organs revealed in each instance a healthy condition. It was noticed that several round-worms were present in the intestines of Pig No. 1. They were not found in Pig No. 4. The organs of these pigs were then subjected to a most careful analysis, conducted quantitatively, to determine the presence or absence of aluminum. The method used throughout all the work by the writer was to ash the material, dissolve in hydrochloric acid, separate the iron by pouring into hot sodium hydroxide solution, and after acidifying with hydrochloric acid precipitating with ammonia. The method yielded quantitative results with known amounts of aluminum and detected weighable amounts. Even in the absence of aluminum it was common to obtain residues weighing 0.1 or 0.2 mg. of a dark reddish yellow color. The qualitative test for the presence of aluminum was the production of a blue color with cobalt nitrate. During these chemical analyses great care was used to prevent the introduction of aluminum, both by the use of pure chemicals, previously tested as to the presence of aluminum, and of platinum and nickel dishes. The chemicals were tested by use in analysis of specimens of milk and of some

of the organs of Pig Nos. 1. The organs of Pig No. 4 examined were:
Gluteal muscle, psoas muscle, liver, kidney, spleen, heart, blood,
brain, thyroid gland, thymus gland, spinal cord, femur (bone)
and bodies of the vertebrae. The amounts taken varied from 50 to
100 gm., unless the entire organ weighed less than this amount, in
which case the entire organ was used; with bone, 10 gm. were taken.
In no instance could aluminum be detected.

EXPERIMENTS WITH MEN[*,2]

Subjects: R., Male, aged twenty-eight; S., Male, aged thirty-four.
Examination as to physical condition and general health led to the
opinion that the subjects were normal. Duration of experiment:
October 18, 1900, to March 17, 1901 (R) and March 14, 1901 (S).
October 19, 1900. Test breakfast consisting of the following:
Baking powder bread (Parrot and Monkey powder) 60 gm.
Water 350 cc.

In one hour after the completion of the breakfasts the stomach
contents were withdrawn through a stomach tube, without the
use of water. Examination of the material revealed the presence of
a liquid together with undissolved sediment. The microscopic
examination of the sediment showed only such food residue as is
derived from the bread taken in the test breakfast. There was
nothing seen that in any way suggested pathological conditions.
The liquid portion, separated by filtration, yielded by analysis the
following results:

TABLE XXVII

	R Per Cent	S Per Cent
Total acidity, as HCl	0.197	0.170
Free hydrochloric acid	0.112	0.088
Combined hydrochloric acid	0.081	0.083
Total hydrochloric acid	0.193	0.171
Acid salts and organic acids	0.004	0.000

October 19, after test breakfast, to October 22, supper, free and
unrestricted diet. Supper, October 22, consisted only of 3 glasses of
milk together with 4 capsules of lampblack for each subject.
October 23 and 24, subjects' diets were as follows: Three meals
each day, each meal consisting of (R) 150 gm., (S) 200 gm. of baking

* January 11, 1902.

powder bread and (R) 1 liter, (S) 750 cc. of milk, the total ingesta during the two days being; baking powder bread (R) 900 gm., (S) 1200 gm., milk (R) 6, (S) 4½ liters. Breakfast, October 25, consisted only of 3 glasses of milk together with 4 capsules of lampblack. The bread was made according to the following recipe:

Flour...................................... 1000 gm.
Lard...................................... 40 gm.
Baking powder (Parrot and Monkey).... 27 gm.
Salt...................................... 12 gm.
Water...................................... 700 cc.

Directions: Sift together the dry ingredients three times, work in the lard with the hands and stir in the water. Bake in a hot oven in two loaves.

Analysis showed that the baking powder contained 6.81 per cent aluminum, as Al_2O_3.

The crumb only was used for consumption and was prepared by breaking into small bits, the size of a marble or smaller, thoroughly mixing the pieces and immediately weighing enough separate portions for each meal during the entire time (two days) and wrapping in paper for preservation. At the same time a portion was taken for analysis, which showed the following composition:

TABLE XXVIII

	Per Cent
Water	44.80
Solids	55.20
Proteins	7.31
Fat	2.29
Carbohydrates	43.92
Inorganic salts	1.68
Aluminum oxide	0.106

Analysis of a composite sample of the milk for the two days showed the following composition:

TABLE XXIX

	Per Cent
Water	87.13
Solids	12.87
Proteins	3.47
Fat	4.02
Carbohydrates	4.61
Inorganic salts	0.77

The total nutrients ingested by the subjects on October 23 and 24 were therefore as follows, the amount being expressed in grams:

TABLE XXX

	Solids	Pro-teins	Fat	Carbohy-drates	Inor-ganic salts	Alumi-num oxide
R						
Bread.............	496.8	65.79	20.61	395.28	15.12	0.954
Milk..............	772.2	208.2	241.2	276.6	46.2	
Total.............	1269.0	273.99	261.81	671.88	61.32	0.954
S						
Bread.............	662.4	87.72	27.48	527.04	20.16	1.272
Milk..............	579.15	156.15	180.90	207.45	34.65	
Total.............	1241.55	243.87	208.38	734.49	54.81	1.272

During the time the subjects were on this known diet, namely October 23 and 24, and for some days following as well, all the feces were collected and subjected to careful examination. It was found that the lampblack and milk ingested on October 22 for supper and October 25 for breakfast, being the food eaten immediately preceding and immediately following the two days on which the subjects were on the known diet, gave such a consistency and color to the fecal matter as to render possible the separation of the feces the formation of which corresponded to this special period. This fecal matter was dried and found to consist of (R) 66 gm., (s) 59.9 gm. of "air dry" material. Chemical analysis showed the following composition:

TABLE XXXI

	R Per Cent	S Per Cent
Water.....................	7.78	13.57
Solids....................	92.22	86.43
Proteins.................	20.59	26.09
Fat......................	28.93	21.55
Carbohydrates...........	6.15	10.33
Inorganic salts...........	36.55	28.46

The total constituents of the feces for this period were therefore, in grams:

TABLE XXXII

Solids	Proteins	Fat	Carbo-hydrates	Inorganic salts
(R)60.87	13.59	19.09	4.06	24.12
(S)51.77	15.63	12.90	6.19	17.05

Deducting these amounts from the nutrients ingested during this period, the following amounts of nutrients remained, representing what was utilized by the body:

TABLE XXXIII

Solids	Proteins	Fat	Carbohy-drates	Inorganic salts
		Totals, grams		
(R)1208.13	260.40	242.72	667.82	37.20
(S)1189.78	228.24	195.47	728.30	37.76
		Per cent of total ingesta utilized		
(R) 95.2	95.0	92.7	99.4	60.7
(S) 95.8	93.8	93.8	99.1	68.9

At 7 A.M. on October 24 the urinary bladder was emptied and the urine passed thereafter for twenty-four hours was collected. Analysis of this urine yielded the following results, (grams):

TABLE XXXIV

	R	S
Volume, cc.	955	788
Specific gravity.	1.026½	1.031
Total nitrogen.	18.002	15.153
Equivalent to urea.	38.487	32.473
Uric acid.	0.383	0.4105
Total phosphate (P_2O_5).	2.789	2.585
Indican.	Moderate amount	Moderately large amount
Total sulphates (H_2SO_4).	4.671	4.670
1, Performed sulphates.	4.535	4.558
2, Combined sulphates.	0.136	0.112

From October 25, 1900 until (R) March 14, (S) March 11, 1901, the subjects partook of a diet which was free and unrestricted excepting as regards the character of the bread and pastry. These consisted entirely of bread and pastry made with Parrot and Monkey brand baking powder. A considerable amount of such food was eaten with each meal and the total was much in excess of the baking powder food that an individual would consume under ordinary circumstances.

On March 2 (R) and February 28 (S) 1901, the subjects were subjected to medical examinations. Physical conditions, normal. No food having been eaten since dinner in the late afternoon of the previous day, the following test breakfast was eaten:

Bread (P. and M.)...................... 60 gm.
Water................................. 350 cc.

After one hour the stomach contents were removed through a tube without the use of water.

Examination of the material revealed the presence of a liquid portion containing in suspension undissolved residue of undigested and partially digested bread. No other food residue was found nor anything seen which in any way suggested a pathological process or condition in the stomach. On analysis the liquid portions yielded the following results:

TABLE XXXV

	R, Per Cent	S, Per Cent
Total acidity, as HCl	0.239	0.311
Free hydrochloric acid	0.141	0.116
Combined hydrochloric acid	0.089	0.176
Total hydrochloric acid	0.230	0.292
Acid salts and organic acids	0.009	0.019
Aluminum oxide	0.0017	0.0017

On March 7, 1901, the subject (S) was again subjected to examination by means of a test breakfast, identical in composition with that of February 28; as on previous occasions, nothing was eaten after the meal of the previous night. In exactly one hour the stomach contents were removed through a stomach tube, without the use of water, and this material subjected to analysis. The microscopic examination of the sediment showed only food residue from the test breakfast. Nothing was seen to suggest a pathological process. The liquid portion yielded the following results of analysis:

TABLE XXXVI

	Per Cent
Total acidity, as HCl	0.317
Free hydrochloric acid	0.131
Combined hydrochloric acid	0.175
Total hydrochloric acid	0.306
Acid salts and organic salts	0.011

With subject R on March 15 and 16 and with Subject S on March 12 and 13, determinations of the utilization of a known diet were made similarly to the determinations made in October. Bread made by the same recipe and employing the same baking powder were used; and in other respects exactly duplicate procedures were followed. The bread crumb presented the following composition:

TABLE XXXVII

	Per Cent
Water	41.58
Solids	58.42
Proteins	8.71
Fat	2.57
Carbohydrates	45.35
Inorganic salts	1.79
Aluminun oxide	0.104

Analysis of composite samples of the milk consumed by (R) and (S) during the two days, showed the following compositions:

TABLE XXXVIII

	R Per Cent	S Per Cent
Water	87.60	86.33
Solids	12.40	13.67
Proteins	3.41	3.41
Fat	3.66	4.76
Carbohydrates	4.54	4.72
Inorganic salts	0.79	0.78

The total nutrients ingested by the subjects on the respective two days were therefore as follows, the figures expressing grams:

TABLE XXXIX

	Solids	Proteins	Fats	Carbohydrates	Inorganic salts
R					
Bread...................	525.78	78.39	23.13	408.15	16.11
Milk...................	744.0	204.6	219.6	272.4	47.4
Total.................	1269.78	282.99	242.73	680.55	63.51
S					
Bread...................	701.04	104.52	30.84	544.2	21.48
Milk...................	615.15	153.45	214.20	212.4	35.10
Total.................	1316.19	257.97	245.04	756.6	56.58

The feces corresponding to the two days on known diet were marked and separated as described for the similar test in October. The amount thus obtained, when it had been reduced to an "air dry" condition, by drying at 65°C. was 64.3 gm. with (R) and 71 gm. with (s). Chemical analysis showed the following composition:

TABLE XL

	R Per Cent	S Per Cent
Water........................	6.19	11.88
Solids........................	93.81	88.12
Proteins.....................	17.84	25.00
Fat.........................	22.27	31.03
Carbohydrates...............	16.84	6.88
Inorganic salts...............	36.86	25.21

The total constituents of the feces of this period were therefore, in grams:

TABLE XLI

Solids	Proteins	Fat	Carbohydrates	Inorganic salts
(R)60.32	11.47	14.32	10.83	23.70
(s)62.57	17.75	22.03	4.89	17.90

Deducting the amounts of these nutrients lost in the feces from the total nutrients ingested during this period, the following amounts remain, which represent the nutrients utilized by the body:

TABLE XLII

Solids	Proteins	Fat	Carbohydrates	Inorganic salts
		Totals, grams		
(R)1209.46	271.52	228.41	669.72	39.81
(s)1253.62	240.22	223.01	751.71	38.68
	Per cent	of total ingesta	utilized	
(R) 95.2	95.9	94.1	98.4	62.7
(s) 95.2	93.1	91.0	99.3	68.3

On March 16 to 17 (R), 13 to 14 (s), the twenty-four hours' urines were collected, analysis of which yielded the following results (grams):

TABLE XLIII

	R	S
Volume, cc..............................	1900	870
Specific gravity.........................	1.018½	1.036
Total nitrogen..........................	21.66	18.449
Equivalent to urea......................	45.98	39.611
Uric acid...............................	0.488	0.525
Total phosphates (P_2O_5)...............	3.80	2.784
Indican................................	Trace	Small amount
Total sulphates (H_2SO_4)...............	6.173	4.822
1. Preformed sulphates................	6.023	4.718
2. Combined sulphates................	0.150	0.104
Aluminum oxide........................	None	

SUMMARY AND CONCLUSIONS[2]

Influence of baking powder food upon growth and the general well-being of animals.

A comparison of the weight and healthfulness of the white rats fed on baking powder food, of Experiment 1, shows that the control rats were slightly heavier. This difference was due to one control rat which was unusually heavy. The important point to be noted is that the growth and increase in weight was constant in each group, an indication that the animals of the two groups were equally healthy. Nothing was observed to indicate or suggest that there was any preference in the two diets.

The increase in weights of the control and baking powder pigs was also very similar. While the pig is a hardy animal it is to be remembered that the process of digestion closely resembles that in man and that the bread consumed, which formed such a large portion of the diet, was made with double the quantity of baking powder recommended for use. Notwithstanding these facts, there were absolutely no indications of disturbances of function or general health.

The experiments then do not show a deleterious action of baking powder food on the growth and healthfulness of animals, even when it is used continuously as the chief article of food and contains double the needed amount of baking powder. Even under the rigid and unnatural conditions of this test the food was consumed without apparent injurious effect.

Influence of baking powder food when used during long periods of time, upon the health of human subjects.

The effect of the five months' feeding with baking powder food in the experiments with men was determined by inquiry along the following lines: 1. Influence on digestion as shown by (a) Gastric secretion; (b) Utilization of known dietary; (c) Absorption and elimination through the urine of products of intestinal fermentation and putrefaction. 2. Influence on metabolism as indicated by the waste products eliminated by the urine when the subjects were on a fixed diet. 3. Influence on general health as indicated objectively and by subjective symptoms.

(a) Reviewing the results on gastric secretion, as indicated by the composition of the stomach contents after a test breakfast, the following will be noted:

	R*		S	
	Beginning per cent	End, per cent	Beginning, per cent	End, per cent
Total hydrochloric acidity...	0.193	0.230	0.171	0.292 0.306

* Previous observation before baking powder bread ingestion: Total hydrochloric acidity 0.220 per cent (see p. 62).

It is thus seen that there was an increased acidity of the stomach contents at the end of the experimental period. The reason of this is not certain. Is it due to nervous influence, the beginning observations being made at a favorable time of the year early in the season's work and when the subjects were in a refreshed and rested condition while the end observations were made after five months of work with its attending fatigue; or is it due to the effect of the continuous eating of carbohydrate food, which was an essential part of the experiment; or is it due to the action of the trace of aluminum which we have seen was dissolved in the stomach contents? These two observations are quite insufficient to determine this point. It is quite unfair to conclude that it was due to any one of these factors, and we are not justified in concluding more than the fact that this slight increase existed. The only subjective manifestation which it produced was a slight desire to drink water, in the case of Subject s.

(b) Bringing together the utilization figures we have the following:

TABLE XLIV

SUBJECT R

		Total solids, per cent	Proteins, per cent	Fat, per cent	Carbohydrates, per cent
Total ingesta utilized.....	Beginning	95.2	95.0	92.7	99.4
Total ingesta utilized.....	End	95.2	95.9	94.1	98.4

SUBJECT S

		Total solids, per cent	Proteins, per cent	Fat, per cent	Carbohydrates, per cent
Total ingesta utilized.....	Beginning	95.8	93.8	93.8	99.1
Total ingesta utilized.....	End	95.2	93.1	91.0	99.3

From these it appears that utilization was practically the same at the beginning and at the end of the observation periods. The variations, one way or the other, are not greater than is expected in parallel observations.

(c) The indications from the urine of toxemia of intestinal origin at the beginning and at the end of the experimental period are as follows:

TABLE XLV

Subject R		Subject S	
Beginning	End	Beginning	End
Indican Moderate amount	Trace	Moderately large amount	Small amount
Combined sulphates 0.136 gm.	0.150 gm.	0.112 gm.	0.104 gm.

The inference from these results (diminished indican) is that there was somewhat less toxemia at the end than at the beginning of the experimental periods. The explanation for this difference may be found in the increased gastric acidity which diminished bacterial activity in the upper intestine by rendering the contents less alkaline; or in the continued ingestion of carbohydrate food.

2. In the experiments on men, by the study of the urine on the several days of the fixed diet at the beginning and end of the experimental periods, we have data which give an indication of certain of the metabolic processes and of the renal function. These data are the volume, specific gravity, nitrogen, uric acid, total phosphates and total sulphates. The results were (grams):

TABLE XLVI

SUBJECT R

	Beginning	End
Volume cc.	955	1900
Specific gravity	1.026½	1.018₅
Total nitrogen	18.002	21.66
Equivalent to urea	38.487	45.98
Uric acid	0.383	0.488
Ratio urea and uric acid	100.7	94.1
Total phosphate	2.789	3.80
Ratio urea and total phosphates	13.5	12.1
Total sulphate	4.671	6.173
Ratio urea and total sulphates	8.2	7.4

TABLE XLVI. (*Continued*)

SUBJECT S

	Beginning	End
Volume cc....................................	788	870
Specific gravity.............................	1.031	1.036
Total nitrogen..............................	15.153	18.449
Equivalent to urea..........................	32.473	39.611
Uric acid...................................	0.4105	0.525
Ratio urea and uric acid....................	79.1	86.8
Total phosphates...........................	2.585	2.784
Ratio urea and total phosphates.............	12.6	14.2
Total sulphates.............................	4.670	4.822
Ratio urea and total sulphate...............	7	8.2

The most striking result in the urine of R is the increase of about 50 per cent in the amount of sulphates. The explanation is that the diet contained sulphates from the baking powder used, which have been absorbed and eliminated by the kidneys. Their presence is, therefore, not indicative of a disorder of any kind. Another noticeable fact, in both Subjects R and S, is an increase at the end of the periods of the quantities of nitrogen, phosphate and sulphate eliminated. This is undoubtedly due to the fact that the amounts of the nitrogenous constituents of the ingested material which were stored in the body, presumably constructed into body tissue, were greater at the beginning than at the end of the experiments, so that conversely, the amount of nitrogenous material undergoing combustion and forming nitrogenous waste products was less at the beginning and greater at the end of the experimental periods. These facts are brought out by the following figures:

TABLE XLVII

	Beginning, gm.	End, gm.
R		
Nitrogen of utilized ingesta.....................	20.83	21.77
Eliminated nitrogen...........................	18.02	21.66
S		
Nitrogen of utilized ingesta.....................	18.26	19.21
Eliminated nitrogen...........................	15.15	18.45

In other words, in the observation at the beginning the body was gaining protein, while at the end the body was in nitrogenous equilibrium in the case of R, and nearly so in the case of S. This explains why the urinary products of destructive metabolism, namely the nitrogenous products, the phosphates and the sulphates, were greater in quantity at the end. It seems that the bodies were in a better nourished condition at the end of the periods in which baking powder was consumed than at the beginning.

The quantity of uric acid eliminated is quite low in all the specimens, as is to be expected from the character of the diet, which indicates that only endogenous uric acid was eliminated, and gives no indications of disorders of nutrition.

It is noticeable that in Subject S the specific gravity is high while the volume is correspondingly small. This is apparently due to the limited amount of water ingested, which consisted only of that in the bread and milk. The condition is not indicative of disorder of any kind. In Subject R at the beginning, the urine presented to a less degree the same character as that noted in Subject S. At the end, the volume was nearly doubled and the specific gravity correspondingly low. The diet of R contained more water than that of S, explaining why in the former the specific gravity was in general lower; but no explanation is apparent for the quite marked difference between the beginning and the end in Subject R. There appears to be no reason to believe that this is indicative either of improved or impaired healthfulness of any of the body organs or of the body as a whole.

3. During the course of the feeding experiments with men careful observation was made of the subjects to determine whether or not there were any objective evidences of disturbances of health. Careful inquiry was also made to determine whether there were subjective manifestations in any way unusual. It was found that Subject R remained practically stationary in weight while Subject S gained about 10 pounds. At no time was any disturbance of digestion or nutrition apparent and the subjects stated at all times that they felt perfectly well and in their usual health. During the whole period the bowels were regular and there was no sensation

of oppression experienced in connection with digestion. In fact, it appeared that there was no abnormality of any kind relative to health, general feeling or nutrition produced by the bread and other food prepared with baking powder.

What becomes of the alumina residue in the food? The evidence of these observations is that such residue passes through the alimentary tract unabsorbed and is evacuated in the stools. In the pig experiments the amounts of baking powder consumed were enormous compared with what man would consume under ordinary conditions. Yet in this animal, whose digestive apparatus closely resembles that of man, there was no retention of aluminum in any of the body organs, not even in small amounts.

Likewise the examination of the urine of Subject R at the end of the long period of subsistence on baking powder food failed to show the presence of aluminum.

In conclusion, the evidence of these experiments is:

1. That the residue of alumina in baking powder food is not absorbed or stored in the body.

2. That such food does not interfere with the growth and well-being of lower animals.

3. That it is not injurious to man even when consumed in large amounts for a considerable length of time.

INVESTIGATION ON THE EXCRETION IN THE FECES OF ALUMINUM INGESTED WITH FOOD PREPARED WITH A SO-CALLED ALUM BAKING POWDER.[*,2]

The following experimental observations were made to ascertain the extent to which aluminum, contained in the residue left in the food prepared with a so-called alum baking powder, is eliminated in the feces.

The subjects of the experiment were four:
  I. Man, young adult.
  II. Man, young adult.
  III. Woman, young adult.
  IV. Child, boy of twelve years.

[*] March 12, 1903.

On November 24, 1902, the subjects were put on a special diet, selected to be as free as possible from aluminum. As will be seen, the traces of aluminum mechanically adherent to the food as soil and from other sources, although minute in amount, being inappreciable in the flour and food stuffs examined, were sufficient to introduce constantly into the feces appreciable amounts of this substance. Accordingly, the amounts present in the feces were determined before the experiment and deducted from the aluminum introduced in the bread during the experimental period. The amounts of aluminum found in the fresh feces in the fore period were as follows (results expressed as per cent of Al_2O_3):

TABLE XLVIII

	Per Cent
Subject I	0.0270
Subject II	0.0236
Subject III	0.0065
Subject IV	0.0204

The baking powder food to be tested consisted of bread made after the following recipe:

TABLE XLIX

	Gm.
Flour	1500
Water	1000
Baking powder (Parrot and Monkey)	40
Salt	18
Lard	40

The dry ingredients were sifted together, the lard well worked into the mixture, the water added and the entire material wet up by hand. It was baked in an ordinary kitchen range oven in the way customary in domestic practice. When the bread was a day old it was put through a chopping machine and the crumb spread in pans and allowed to dry in the air. This is the crumb fed in the experiments. It was bottled, sampled and analyzed, the amount of aluminum being 0.204 per cent Al_2O_3.

During the two days December 16 and 17, a considerable quantity of this bread was added to the dietary, other food of the special diet being eaten without restriction as to quantity. The amount of bread eaten by each of the subjects and the amount of aluminum accordingly ingested were as follows:

TABLE L

	Bread, Gm.	Per Cent	Al_2O_3, Gm.
Subject I	850	× 0.204 =	1.734
Subject II	750	× 0.204 =	1.530
Subject III	300	× 0.204 =	0.612
Subject IV	375	× 0.204 =	0.765

The feces corresponding to this period were marked by the ingestion of carmine with the first and last meal. At the first appearance of carmine in the stools they were saved as were all subsequently passed, until the stools were again colored and free from carmine coloration.

This material was dried by evaporation with addition of small quantities of alcohol, powdered and analyzed. The total amounts of powdered material and amounts of aluminum were as follows:

TABLE LI

	Dry feces	Per cent Al_2O_3	Total Al_2O_3
Subject I	111.5	1.634	1.817
Subject II	109.7	1.455	1.590
Subject III	51.4	1.188	0.612
Subject IV	79.3	1.007	0.801

There remains to deduct from these figures the small amounts calculated to have been introduced with the food as mechanical impurities. They were calculated to be as follows:

TABLE LII

	Moist feces	Per cent Al_2O_3 without bread	Total, gm.
Subject I	222.0	0.0270	0.060
Subject II	225.7	0.0236	0.051
Subject III	223.9	0.0065	0.014
Subject IV	176	0.0204	0.036

Deducting the amounts found in the feces:

TABLE LIII

	Total in feces from bread, gm.		Total in bread, gm.	Per cent recovered
Subject I.................	1.817–0.060	1.757	1.734	101.3
Subject II................	1.590–0.051	1.539	1.530	100.6
Subject III...............	0.612–0.014	0.598	0.612	97.7
Subject IV...............	0.801–0.036	0.765	0.765	100
			Average	99.9

It is seen from these figures that the entire amount of the aluminum from the baking powder ingested with the food was present in the feces and it followed that this was the channel of elimination from the body of this constituent of the baking powder residue.

VI.

THE INFLUENCE OF ALUMINUM COMPOUNDS ON THE NUTRITION AND HEALTH OF MAN

A Digest of the Report[1] of the Referee Board of Consulting Scientific Experts, Created by Executive Order of President Roosevelt in February, 1908

PERSONNEL OF BOARD

The members of the Referee Board were selected by President Roosevelt after correspondence with the presidents of the leading universities of the country:

Chairman, IRA REMSEN,[3] M.D., PH.D., LL.D., D.C.L., professor of chemistry and president of Johns Hopkins University; Fellow of the National Academy of Sciences (president 1907–1913); member, American Chemical Society (president, 1902); Fellow, Society of Chemical Industry (president 1909); American Association for the Advancement of Science (president 1903); American Academy of Arts and Sciences; Philosophical Society; corresponding member, British Association for the Advancement of Science; foreign member, London Chemical Society; honorary member, Pharmaceutical Society of Great Britain; foreign honorary member, Société Chemique de France.

RUSSELL H. CHITTENDEN,[3] PH.D., SC.D., LL.D., professor of physiological chemistry, Yale University, and Director of the Sheffield Scientific School; Fellow of the National Academy of Sciences; member, Society of Naturalists (president 1893); American Physiological Society (president 1895–1904); Society of Biological Chemists (president 1907); Fellow, Philosophical Society; American Academy of Sciences; associate member, Société Royale des Sciences Médicales et Naturelles de Bruxelles.

JOHN H. LONG,[2] SC.D., professor of chemistry, Northwestern University Medical School; chemist, Illinois State Board of Health, 1885–1905; member of the Committee on Revision, U. S. Pharmacopoeia; Fellow, American Association for the Advancement of Science (president, 1903); member, Society of Biological Chemists; Washington Academy of Sciences; Deutsche Chemische Gesellschaft.

ALONZO E. TAYLOR,[3] M.D., Benjamin Rush professor of physiological chemistry, University of Pennsylvania; member of the Society of Naturalists; American College of Physicians; Society of Biological Chemists; Society of Experimental Biology; Deutsche Chemische Gesellschaft.

THEOBALD SMITH,[3] A.M., M.D., LL.D., professor of comparative pathology, Harvard University (since 1915, director, Department of Animal Pathology, Rockefeller Institute); (appointed on the Referee Board to succeed Prof. Christian A. Herter, who died in 1910); Fellow, National Academy of Sciences; member, Society of Naturalists; Fellow, Philosophical Society; member, Society of Experimental Pathology; Society of Pharmacology; Association of Pathologists and Bacteriologists; Fellow, American College of Physicians; Member, Society of Bacteriologists; Society of Experimental Biology; Fellow, American Academy of Science; honorary Fellow of the London Society of Tropical Medicine and Hygiene; honorary member, Société de Pathologie exotique.

THE PROBLEM AND CONCLUSIONS

"The questions bearing upon the influence of aluminum compounds on the health of man referred to this Board were as follows:

"'1. Do aluminum compounds, when used in foods, affect injuriously the nutritive value of such foods or render them injurious to health?

"'2. Does a food to which aluminum compounds have been added contain any added poisonous or other added deleterious ingredient which may render the said food injurious to health?

"'(a) In large quantities.

"'(b) In small quantities.

"'3. If aluminum compounds be mixed or packed with a food, is the quality or strength of said food thereby reduced, lowered, or injuriously affected?

"'(a) In large quantities.

"'(b) In small quantities.'

"By the term 'small quantity' we understand such an amount as may be ingested in the occasional use of biscuits, pastry, or other articles leavened with baking powder, as these foods are practically used in the ordinary American family. This amount will not average more than 25 mg. to 75 mg. of aluminum daily, for the days of consumption of such articles.*

"By the term 'large quantity' we understand such an amount of aluminum as would be ingested only under very unusual conditions, as, for example, where the flour consumption is mainly in the form of biscuits or other articles leavened with aluminum baking powders. This amount may reach 150 mg. to 200 mg. of aluminum per day. A person subsisting mainly on baking powder biscuits, as may happen in camp life, might ingest an amount in excess of 200 mg. per day. With this possibility in mind, we have also studied the effects of amounts up to and exceeding 1000 mg. of aluminum per day.

"Our answers to the questions submitted are as follows:

"1. Aluminum compounds when used in the form of baking powders in foods have not been found to affect injuriously the nutritive value of such foods.

"2. (a) Aluminum compounds when added to foods in the form of baking powders, in small quantities, have not been found to contribute any poisonous or other deleterious effect which may render the said food injurious to health. The same holds true for the amount of aluminum which may be included in the ordinary consumption of aluminum baking powders furnishing up to 150 mg. of aluminum daily.

"(b) Aluminum compounds when added to foods, in the form of baking powders, in large quantities (up to 200 mg. or more per day) may provoke mild catharsis.

*For the author's definition of the R. B. unit, see p. 326.

"(c) Very large quantities of aluminum taken with foods in the form of baking powders usually provoke catharsis. This action of aluminum baking powders is due to the sodium sulphate* which results from the reaction.

"(d) The aluminum itself has not been found to exert any deleterious action injurious to health, beyond the production of occasional colic when very large amounts have been ingested.

"3. When aluminum compounds are mixed or packed with a food, the quality or strength of said food has not been found to be thereby reduced, lowered, or injuriously affected.

(*Signed*)   IRA REMSEN, *Chairman*

RUSSELL H. CHITTENDEN

JOHN H. LONG

ALONZO E. TAYLOR

THEOBALD SMITH

Referee Board of Consulting Scientific Experts"

1. EXPERIMENTAL STUDY BY RUSSELL H. CHITTENDEN

The experiments were made on twelve men and continued from January 15 to June 22, 1912. During one hundred and thirty days the diet contained bread raised with an alum baking powder (test bread No. 1) made in the laboratory.

This bread was made fresh every day and contained in one baking (two loaves) approximately:

TABLE LIV

		gm.
Sifted flour................................... qt. 2		1080
Baking powder,† heaping teaspoonfuls...........	4	40
*Salt (approximately one rounded teaspoonful).... oz. ⅓		10
Butter....................................... oz. 1		30
Water, sufficient quantity.		

* "The aluminum compound made use of in our experimental work which is the compound ordinarily employed in the manufacture of alum baking powders was sodium aluminum sulphate, $NaAl(SO_4)_2$."

	gm.
† Sodium aluminum sulphate, calcined................	25
Sodium bicarbonate, dry...........................	26
Corn starch..	49
	100

Bread free from the aluminum compounds was raised by yeast. Sodium sulphate yeast bread (test bread No. 2), was made by incorporating sodium sulphate in the yeast bread:

The dose of aluminum compound was increased from time to time, at first by increasing the quantity of bread, and later by increasing the quantity of the baking powder used in making the bread. In this way the alum used per man per day was increased from 0.578 gm. in the beginning to 2.314 gm. in later dosage periods; the actual aluminum contained in this dosage ranged from 0.065 gm. to 0.260 gm. per man per day. Eight men used the alum bread, while 4 had no added aluminum in their food, but 2 of them were given sodium sulphate (test bread No. 2) during the later periods of the experiment.

On the basis of six days' preliminary observations to determine the amount of food desired by each individual, a definite menu was arranged for the different meals of each day, sufficiently generous in character to admit of reasonable choice without being unduly large, and enabling each individual to exercise to some degree his personal likes and dislikes, yet keeping the daily variety within reasonable limits.

On the basis of the diets of all the subjects during three of the ten-day periods, it was estimated that the fuel value of the diet averaged about 3000 large calories per day, although the extreme variations on individual days were 1950 and 4925 calories.

In the daily average of the individual periods estimated, the extreme variations were 2269 and 3592 calories.

The daily intake of nitrogen was accurately determined and varied with all subjects, in ten-day period averages, from 10.91 gm. to 23.09 gm. These differences were due to the variation in diets of the individual subjects; in the first ten of the twenty days' fore-period the subjects consumed daily a larger amount of nitrogenous food; subsequently throughout the experimental periods, the subjects maintained within narrow limits their individual nitrogen intake levels.

Albumen was present in traces in the urine at rare intervals, but no more frequently in the aluminum than in the control periods and not in more of the aluminum than of the control subjects. Casts appeared in the urine of only one subject and then in the early control period disappearing in the course of aluminum ingestion.

R. F. Rand, M.D., the medical examiner in charge, reported:

"The men presented considerable variation in age, general physique and habits of exercise, etc. They showed a striking uniformity in the fact that they went through the six months of the test with few disturbances of health and these of a minor nature.

"I could find no evidence that the feeding with alum in any way affected their general health, their nutrition, or digestion, or had any effect on the excretory functions of the kidneys or intestines."

There was a slight loss of *body weight* during the experiment by the four controls and 3 aluminum subjects, practically no change with 3 and slight gain with 2 aluminum subjects.

The *blood* hemoglobin and morphology in general showed only such variations as occur in average individuals over a similar length of time. With the individual subjects, the greatest differences in the average hemoglobin figures for periods were:

Experimental subjects, a loss of 8 points and a gain of 4 points; average of all subjects, a loss of 2 points. Control subjects, a loss of 12 points and a gain of 5 points; average of all subjects a loss of 4 points.

There was a slight increase in the weight of *feces* during the later periods of the experiment, when aluminum salts in relatively large amounts were being taken with the food, but diminishing in the after ten-day control period in all of the subjects but one. Similar changes in the feces were noted in one of the sodium sulphate controls and also appeared in one of the two other controls. While in some instances the increase was accompanied by an increase in the amount of moisture, it was in no instance greater than appeared in one of the controls without sodium sulphate. Therefore little if any significance is attached to the relatively small difference, although on general principles, bearing in mind the well-known action of sodium sulphate, some slight effect on the amount of the feces and their water content might be expected.

Studies of the fecal bacteria included total count (Eberle-Klein method), viable (agar plates) Gram+ and Gram–, Gram+ and Gram– bacilli and cocci, spores and gas formation by dextrose-broth fermentation. There were variations in the data obtained but they bore no relationship to the ingestion of aluminum compounds or of sodium sulphate.

Tests of the feces for phenols, indol, skatol and hydrogen sulphide showed no differences as to the presence of these bacterial decomposition products following the ingestion of the aluminum compounds or of sodium sulphate.

The *fat utilization*, as determined from comparison of the fat of the food, accurately determined throughout the experiment, and of the feces (ether extract), uniformly varied from 96 to 98 per cent, even when the diet was relatively rich in fat and when aluminum was present in the food.

The utilization of protein food, as determined by the nitrogen balance of the food and feces, varied from 88 to 93 per cent and when aluminum was present showed a diminution that amounted to 1½ per cent in three subjects with practically no diminution in the other five. One of the controls likewise showed a diminution of 1½ per cent during the same period.

The feces, after two days of special diet, were examined by the Schmidt-Steele method and also for hydrobilirubin. There were no differences during the aluminum periods.

The *urine* showed no significant variations in volume and specific gravity during the aluminum periods; or in total nitrogen or phosphate-phosphorus; or in the ratio of phosphorus to nitrogen, as indicated in the following summary of data of these constituents:

TABLE LV

URINARY PHOSPHORUS AND NITROGEN ELIMINATION
DAILY AVERAGES, GM.

Date		Jan. 15–Feb. 3	Feb. 4–May 3	May 4–13	May 14–23	May 24–June 12	June 13–22
Aluminum ingested, av. mg. per day		0	68	140	186	258	0
J.A.A.	P	1.04	0.97	1.03	1.01	1.11	1.05
	N	13.83	14.15	14.55	13.66	15.12	13.59
	N/P	13.3	14.7	14.1	13.5	13.6	12.9
R.B.H.	P	1.33	1.09	1.04	1.03	1.11	1.20
	N	15.83	14.34	13.41	13.46	14.25	14.71
	N/P	12.0	13.1	12.9	13.0	12.9	12.2
B.I.	P	0.70	0.74	0.89	0.92	1.09	1.14
	N	11.22	11.69	11.47	10.97	13.26	13.81
	N/P	16.1	15.3	12.8	11.9	12.1	12.1
R.H.S.	P	1.00	0.82	0.83	0.81	0.98	1.06
	N	13.71	12.57	12.78	11.73	14.25	15.81
	N/P	13.8	15.4	15.4	14.4	14.5	14.9
C.L.D.	P	1.01	0.88	0.89	0.89	0.93	1.01
	N	14.04	13.07	13.18	12.52	13.60	14.18
	N/P	13.9	14.9	14.8	12.9	14.6	14.0
E.S.W.	P	1.19	1.04	1.08	1.12	1.11	1.15
	N	15.83	14.72	16.38	14.71	15.19	15.30
	N/P	13.3	14.2	15.1	13.1	13.6	13.3
J.C.T.	P	1.33	1.14	1.17	1.18	1.28	1.20
	N	18.24	17.59	17.89	17.51	18.54	17.53
	N/P	13.7	15.4	15.2	14.8	14.7	14.6
C.J.T.	P	1.21	1.08	1.10	1.18	1.26	1.20
	N	17.46	17.12	17.54	18.28	18.99	17.49
	N/P	14.5	15.8	15.9	15.4	15.1	14.5

CONTROLS

H.B.L.	P	0.96	0.89	0.98	0.96	1.12	1.01
	N	14.50	14.95	16.10	15.89	17.21	16.70
	N/P	15.1	17.0	16.4	16.5	15.4	16.5
S.G.	P	1.03	0.91	0.92	0.93	1.02	0.95
	N	14.17	13.17	13.10	12.78	13.35	13.53
	N/P	13.7	14.5	14.2	13.7	13.1	14.2
R.B.W.	P	1.28	1.09	1.16	1.23	1.35	1.26
	N	13.49	13.13	14.40	13.87	14.34	14.30
	N/P	10.5	12.0	12.2	11.2	10.9	11.3
R.L.K.	P	0.98	0.80	0.84	0.84	0.85	0.76
	N	10.53	10.59	10.94	10.89	10.81	9.73
	N/P	10.8	13.2	13.0	13.0	12.8	12.8

Further, there was no evidence of any specific influence exerted by the ingested aluminum upon the character of the nitrogen balance, the subjects throughout the entire experiment being very close to a negative balance.

Tests were made of the urinary phenols and aromatic oxyacids. In 4 of the subjects there was slight reduction in the amount of phenols during the aluminum periods; otherwise there were no differences.

CHITTENDEN'S CONCLUSIONS

"The results of the experiments show that small and relatively large quantities of aluminum compounds, when taken daily with the food, are without effect upon the general health and nutrition of the body as indicated by the subjective and objective signs, by the composition of the urine and feces, the absorption and utilization of the protein and fatty foods, the nitrogen balance, the character of the bacteria of the intestinal tract as indicated by available methods, and the character of the blood as indicated by the hemoglobin content, the number of red and white blood cells, etc.

"In other words, aluminum compounds when used in foods, as in bread, in such quantities as were employed in our experiments, do not affect injuriously the nutritive value of such foods, nor render them injurious to health, so far as any evidence obtained in our experimental work indicates."

II. STUDY ON THE PHYSIOLOGICAL BEHAVIOR OF COMPOUNDS OF ALUMINUM AS USED IN FOODS OF MAN
BY JOHN H. LONG

Six subjects were fed on a "controlled diet" to which were added the products of the reaction of s.a.s. (120 gm. dissolved in water and 126 gm. sodium bicarbonate, heated one hour on the water bath and made up to 3000 cc., 50 cc. being equivalent to 2 gm. of the anhydrous s.a.s.). In the latter half of the work sulphates were removed by washing by decantation. The subjects drank the required amounts of these emulsions, diluted with water or milk, with meals at noon and night. In the following plan of experiments, the dosage is in equivalent of s.a.s. (2 gm. = 0.2236 gm. Al).

TABLE LVI

Periods	Daily dosage, residue gm. S.A.S.	Duration, days
1 to 2. Fore...............................	—	20
3 to 5. Low aluminum.....................	2	30
6. Medium aluminum.................	4	10
7. Rest..............................	—	10
8. Medium..........................	Al(OH)₃ from 4	10
9. Higher...........................	Al(OH)₃ from 6	10
10. Highest..........................	Al(OH)₃ from 10	10
11. Sulphate.........................	Sulphate from 4	10
12. After.............................	.. —	10

The food, urine and ten-day aliquot samples of the feces were analyzed.

The subjects were medical students in good physical condition, but differed greatly in weight (44 lb.) and in robustness and general physical appearance.

Urine. Tests of the urine every five days for albumin, sugar, glycuronic acid, acetoacetic acid and acetone were uniformly negative. The reaction was uniformly acid. The indolacetic acid, aromatic oxyacids, phenol or the sediment showed no possibly significant variations, excepting:

Subject 2. Stronger reactions for indolacetic acid, aromatic oxyacids and in one instance phenol, in Periods 9 and 10.

Subject 6. For a time rather marked quantities of blood and pus in the urine which, however, followed an injection of silver salt into the urethra.

Feces Tests. In general characteristics there were no significant findings. Variations in consistence were from soft to hard without relation to periods, excepting that as a rule they were semisolid in Periods 3 to 6, once being liquid; and in 4 subjects were liquid in Period 11.

Skatol was uniformly absent; indol, phenol and biliary pigment ($HgCl_2$ test) varied without relation to particular periods.

Bacteriological examination of Gram smears suggested a possible relative increase of + bacilli at isolated times in

4 of the subjects. Gram smears from the sediments of plain bouillon and from glucose bouillon growths showed considerable variations.

Blood hemoglobin and morphology showed no significant variations.

The following are the averages of hemoglobin determinations for the particular periods:

TABLE LVII

Subject	1	2	3	4	5	6
Periods						
1–2	91.5	97.2	95.0	92.0	80.2	97.2
3–5	85.6	89.5	83.6	88.3	79.6	88.9
6	91.0	95.0	89.0	91.0	82.5	90.5
7	88.5	89.5	84.0	81.0	81.5	95.0
8	83.0	87.5	78.0	83.5	79.0	85.0
9	94.5	91.5	96.5	96.5	80.5	101.5
10	95.0	95.0	84.5	93.5	85.5	91.5
11	96.0	93.5	86.0	92.0	86.5	93.5
12	95.5	95.0	85.0	85.5	83.5	85.5

On comparison of the figures for the fore-periods (1 and 2) with those of the final aluminum periods (10), it will be observed that variations are within physiological limits and are higher following the ingestion of aluminum compounds in 3 subjects and lower in 3 subjects. An average of all determinations of all subjects, which would tend to eliminate variations either physiological or from technique, gives 90.5 for the fore-periods and 90.8 for the last ten-day period of the seventy-day aluminum administration, in which an amount was ingested probably ten times what would be ingested from baking powder food under unusual conditions and twenty-five times under usual conditions.

Medical Record. Subject 1. Previous history of bringing up blood when clearing throat. May come from teeth or gum. Day 18, symptoms of "grippe." 41, headache; 42, appetite poor, nausea; 55 to 56, abdominal cramps; 58, coughed blood; 61 headache last few days; 73 to 74, few bowel cramps; 74 to 78, tenderness over descending colon; 83, nausea; 93 to 97, cold with bronchitis. Body weight, gain 1.8 kg.

Subject 2. Previous history of abdominal soreness and tenderness. Days 23 to 25, bowel cramps; 37 to 39, slight nausea with bowel cramps; 40, temperature 101.8°F., itches all over, two hard movements; 41, temperature 99.4°F.; 42, temperature 99.2°F.; 43, at night, hard stools with mucus and little blood (external hemorrhoids), next morning diarrhoea; 45, nausea, abdominal pains; 45 to 47, little mucus in stool; 51, pain in region of appendix; 53 to 54, two softer movements daily, cramps, headache; 55 to 57, tenesmus, blood and mucus in stool, frequent movements; 58 to 60, two soft movements daily, symptoms subsiding; 77, 80, 81, pain and tenderness, appendix region; 79, surface of stool blood tinged; 86, pain in bowels with two loose movements; 94, bowel cramps and pain around appendix; 100,107, 108, bowel cramps. Body weight, gain 0.3 kg.

Subject 3. Commonly has gastric distress, three to four hours after eating, with sour eructations, especially when constipated. Day 7, gastric distress and sour eructations, slight constipation; 8, headache, vomited ½ liter; 9, stomach better but not good; 41, slight eczema about arms for several days; 55 to 56, 58 to 59, bowel cramps, somewhat constipated; 56, 74, slight blood in stool. Body weight, gain 2.3 kg.

Subject 4. Days 24 to 25, 55 (58 to 62), 107, 111, bowel cramps. Days not indicated in parentheses, stools soft or liquid. Days 50, 53, 56, 62, 78, slight blood in stool; 98 to 100, cold. Body weight, gain 1.1 kg.

Subject 5. Days 7, 49, 54 to 55, 73, 75, slight blood on toilet paper or stool from fissure of anus; 27, cold for several days; 56 to 57, headache; 59, 98, bowel cramps; 96, 98 to 101, 117, temperature 99 to 99.6°F. Body weight, gain 2.2 kg.

Subject 6. Days 24 to 27, general abdominal tenderness, sore throat; 30, 40, 58, 67, headache; 40, vomited; 38, urethral injection 20 per cent argyrol, red blood corpuscles and leucocytes in urine for following ten days; 7, 26 to 27, 33, 46, 52, 59, 60, 64, 75, 77, looseness of bowels; 59 to 61, cold;

25, 73 to 74, cramps; 100, blood in feces. Body weight, gain 1.0 kg.

Summary of General Medical Condition. "A study of the various tables and charts presented shows that in general the subjects remained in good health throughout the period of the experiment. In the sum of the days of observation there was only one day when one man felt unable to attend his regular college work, and on this occasion the subject was kept at home because of a severe cold.

"It is somewhat significant that no digestive disturbances of any consequence whatever occurred, and the only effects differing from those normally present in all individuals were those of the larger doses of the double sulphate, or its components, on the lower bowel. These effects were anticipated, and the larger doses were administered to see if other disturbing effects, in addition, would follow. But even after the ingestion of the largest amounts of the aluminum compounds there was no loss of appetite, no loss of weight, no effects on the blood or the urinary excretion which pointed to a modification of any of the normal processes.

"It is evident from the experiments that large doses of aluminum salts are liable to be followed by loose movements of the bowels, sometimes accompanied by griping pains, this condition, in turn, being occasionally succeeded by a slight degree of constipation. Such a sequence is frequently observed after the ingestion of repeated doses of sodium sulphate. It will be noted that the behavior of the aluminum sodium sulphate in the large ingestion is frequently duplicated by the corresponding amount of the alkali sulphate, while the largest doses of aluminum hydroxide have, apparently, no more effect than the smaller doses. This is the general situation, but there appear to be occasional exceptions. Taking everything into consideration I am inclined to think that the disturbing action on the bowels of the large ingestions is due to the alkali sulphate by-product, rather than to the alumina fraction.

"It is proper to add, however, that the men themselves were not in any way alarmed by the symptoms shown, as beyond the frequency of the bowel movements there seemed to be no specially characteristic change in the general condition. Other experiences with Subject No. 2 had shown a tendency toward disturbances of the intestinal tract, and this tendency was aggravated by the larger doses of the salt administered."

TABLE LVIII

NITROGEN UTILIZATION PER CENT

Subjects	1	2	3	4	5	6
Periods						
1–2	90.79	89.51	88.48	88.92	90.78	85.94
3–5	86.03	85.33	86.37	85.54	88.78	82.45
6	87.70	85.01	85.45	85.84	88.87	81.13
7	87.34	91.51	87.76	87.13	91.09	83.38
8	87.63	83.99	88.99	86.56	91.01	83.71
9	88.12	86.23	88.34	87.53	90.13	86.53
10	87.59	86.06	88.10	82.06	88.50	84.43
11	90.25	87.34	86.22	85.80	89.60	84.89
12	87.33	81.92	85.65	85.96	84.34	80.50

As a rule, the *nitrogen balances* average better in the dosage than in the control periods. *Nitrogen utilization* is better in the fore-period than in the dosage periods, especially with sulphates. It is low in the after-period following sulphate administration. It is better in the aluminum periods without (8 to 10), than in the aluminum periods with (3 to 6), sulphates.

The *urinary nitrogen distribution* (partition) is remarkably regular throughout.

In Periods 3 to 5 two-thirds or three-fourths of the ingested *sulphur* appears to leave the body in the urine, the remainder by the feces, as inorganic sulphates. With some of the subjects, the ethereal sulphates of the fore periods are higher, possibly because of the laxative action during dosage periods. This is regarded as not objectionable and as tending to prevent the absorption of products of putrefaction.

TABLE LIX

URINARY PHOSPHORUS AND NITROGEN ELIMINATION

Daily Averages, Gm.

Subjects		1	2	3	4	5	6
Periods							
1–2	P	0.73	1.04	0.73	0.77	0.77	0.63
	N	8.99	12.02	9.82	11.10	10.09	7.95
	N/P	12.4	11.6	13.4	14.5	13.2	12.8
3–5	P	0.58	0.77	0.68	0.63	0.69	0.51
	N	7.98	10.00	9.89	10.10	9.88	7.50
	N/P	13.9	12.9	14.6	16.1	14.3	14.6
6	P	0.50	0.69	0.60	0.52	0.70	0.41
	N	8.13	9.94	10.74	9.98	10.23	7.20
	N/P	16.3	14.4	17.9	19.2	14.6	17.6
7	P	0.72	0.85	0.75	0.72	0.88	0.56
	N	9.15	10.60	10.60	10.62	10.46	7.51
	N/P	12.7	12.5	14.1	14.8	11.9	13.4
8	P	0.61	0.75	0.71	0.52	0.71	0.43
	N	9.16	10.26	10.78	9.54	10.37	7.55
	N/P	15.0	13.7	15.2	18.3	14.6	17.6
9	P	0.49	0.68	0.60	0.47	0.59	0.34
	N	9.42	10.96	10.41	9.59	10.06	8.11
	N/P	19.2	16.1	17.4	20.4	17.1	23.9
10	P	0.45	0.71	0.50	0.44	0.52	0.38
	N	9.25	12.14	10.49	8.89	10.01	8.59
	N/P	20.5	17.1	21.0	20.2	19.3	22.6
11	P	0.83	1.01	0.83	0.74	0.82	0.81
	N	8.91	9.97	9.12	7.67	9.33	8.63
	N/P	10.7	9.9	11.0	10.4	11.4	10.7
12	P	0.77	0.93	0.75	0.74	0.73	0.69
	N	9.61	10.46	8.99	8.67	8.69	8.99
	N/P	12.5	11.3	12.0	11.7	11.9	13.0

The *fat utilization* was good throughout with no variations to suggest any influence of aluminum ingestion.

Urinary *indican* is regarded as not related to Al ingestion. The table shows some slight increases, more marked in Period 11 than during aluminum administrations.

Coincident with the decrease of urinary phosphates, urinary *acidity* drops appreciably during Periods 3 to 6, less strikingly during Periods 8 to 10. This is not regarded as objectionable.

TABLE LX

THE ACIDITY OF THE URINE: MEANS FOR ALL SUBJECTS

Subjects	1	2	3	4	5	6
Periods						
1–2	1.37	1.99	1.25	1.54	1.39	1.43
3–5	1.04	1.41	1.02	1.14	1.21	1.01
6	0.89	1.04	0.86	0.96	1.12	0.69
7	1.48	1.59	1.24	1.42	1.65	1.07
8	1.27	1.58	1.31	1.07	1.47	1.07
9	1.25	1.49	1.15	0.99	1.20	0.88
10	1.16	1.61	1.07	0.96	1.36	1.17
11	1.55	2.07	1.49	1.39	1.65	1.82
12	1.82	1.95	1.49	1.37	1.54	1.62

The *dry weights of the feces* show an increase over the fore periods, averaging (3 to 5) 20 per cent; (6) 37 per cent; (7) 25 per cent; (8) 37 per cent; (9) 46 per cent; (10) 48 per cent; (11) 39 per cent; (12) 40 per cent. This is ascribed in part to the ingestion of more vegetables in the later months of the experiments and in part to the residue from the experimental dosage. In Period 12 it is "apparently connected with the poorer food utilization of this period."

FURTHER OBSERVATIONS

Additional observations on subjects not limited in diet and ingesting more nitrogenous food were made on 12 men, divided into three sub-groups of 4 men each. Periods were seven days. Group 1 received daily the residue from 2, 3 and 4 gm. S.A.S., reacting with an equivalent of sodium bicarbonate, prepared as previously described; Group 2, 50 cc. of physiological salt solution and later received daily the residue from the reaction between calcium acid phosphate and $NaHCO_3$, in amounts to yield CO_2 equal to that from 2 gm. S.A.S.; Group 3 received daily the Rochelle salt residue from the reaction between potassium acid tartrate and $NaHCO_3$ in amounts to yield CO_2 equal to that from 2, 3 and 4 gm. S.A.S.

TABLE LXI

URINARY PHOSPHORUS* AND NITROGEN ELIMINATION AND ACIDITY

First Group

Subjects			A	B	C	D
Periods 1	Dosage None	P	0.91	0.76	1.04	0.98
		N	12.38	9.42	17.78	12.72
		N/P	13.6	12.4	17.1	13.0
		AC	2.28	1.74	2.65	2.16
2–4	Res. 2 gm. S.A.S.	P	0.84	0.89	1.10	1.07
		N	11.01	10.73	14.62	12.00
		N/P	13.1	12.1	13.3	11.2
		AC	1.87	1.61	2.35	1.79
5–6	Res. 3 gm. S.A.S.	P	0.95	0.93	1.18	1.04
		N	11.44	12.38	15.70	13.32
		N/P	12.0	13.3	13.3	12.8
		AC	1.70	2.03	2.31	1.47
7–8	Res. 4 gm. S.A.S.	P	0.74	0.78	1.05	0.81
		N	11.12	11.94	14.73	11.25
		N/P	15.0	15.3	14.0	13.9
		AC	1.43	1.46	2.09	1.02
9	None	P	0.68	0.90	1.15	0.78
		N	9.76	12.85	14.94	11.27
		N/P	14.3	14.3	13.0	14.4
		AC	1.68	1.71	2.26	1.35

* P_2O_5 recalculated as P and N/P value calculated, for comparison with previous tables.

TABLE LXI. (*Continued*)
Second Group

Subjects		E	F	G	H
Periods I	Dosage None	P 0.56	0.89	0.95	0.74
		N 9.75	11.70	11.65	10.25
		N/P 17.4	13.2	12.3	13.9
		AC 1.06	1.80	2.37	1.78
2–4	Physiol. Salt Sol.	P 0.60	0.82	1.03	0.78
		N 10.39	10.66	12.42	10.47
		N/P 17.3	13.0	12.1	13.4
		AC 1.13	1.33	2.29	1.50
5–6	Physiol. Salt Sol.	P 0.84	0.77	1.23*	0.80
		N 10.16	10.88	12.74	10.72
		N/P 12.1	14.1	10.3	13.4
		AC 1.09	1.26	2.47	1.40
7–8	Phosphate residue equivalent 2 gm. S.A.S.	P 0.85	1.01	1.23	0.91
		N 9.23	10.80	10.09	10.16
		N/P 10.9	10.7	8.2	11.2
		AC 0.93	1.45	2.09	1.45
9	None	P 0.68	0.83	0.92	0.77
		N 6.14	11.34	11.12	11.08
		N/P 9.0	13.7	12.1	14.4
		AC 1.94	1.48	1.79	1.34

*Subject G. No data on period 6.

In group 1, urinary acidity and phosphates are present to a lesser degree than in the previous observation, somewhat lower during the aluminum dosage, while the nitrogen of the feces shows no definite increase. In 2 of the subjects there is some increase of fecal solids.

In Group 2, there is no lessening of urinary acidity or phosphates, excepting some increase of the latter during Periods 7 and 8 with 2 of the subjects. As a whole, the group shows no marked differences from Group 1.

TABLE LXI. (*Continued*)

Third Group

Subjects			I	J	K	L
Periods I	Dosage None	P	0.73	1.03	0.92	0.75
		N	10.42	12.97	12.83	12.41
		N/P	14.3	12.6	14.0	16.5
		AC	1.83	2.52	2.03	1.83
2–4	Roch. res. equivalent 2 gm. S.A.S.	P	0.82	1.21	1.10	1.01
		N	9.11	12.76	12.93	13.06
		N/P	11.1	10.5	11.8	12.9
		AC	0.92	1.81	1.40	1.16
5–6	Roch. res. equivalent 3 gm. S.A.S.	P	0.87	1.26	1.18	0.98
		N	10.49	11.98	12.78	14.54
		N/P	12.1	9.5	10.8	14.8
		AC	0.62	1.33	1.11	1.05
7–8	Roch. res. equivalent 4 gm. S.A.S.	P	0.68	1.08	1.02	0.93
		N	10.08	12.33	12.62	14.32
		N/P	14.8	11.4	12.4	15.4
		AC	0.25	0.59	0.54	0.74
9	None	P	0.73	1.00	1.12	0.92
		N	11.19	13.24	14.44	14.37
		N/P	15.3	13.2	12.9	15.6
		AC	1.47	1.85	2.12	1.47

In Group 3, there is a definite reduction of urinary acidity. Other factors are negative.

As a whole, the fat of the feces shows no variations attributable to the particular products ingested.

The blood hemoglobin (Dare) and morphology (count and differential), the former observed daily, the latter once during each period, showed no variations attributable to the particular products ingested.

TABLE LXII

BLOOD HEMOGLOBIN

Subject		A	B	C	D	E	F	G	H	I	J	K	L
Period 1	(1)	101	96	100	100	98	98	97	98	99	97	98	100
	(2)	98	94	99	99	95	93	92	94	95	93	95	93
	(3)	100	95	100	100	97	96	95	96	97	95	97	97
2	(1)	100	96	100	100	97	97	96	97	97	97	98	99
	(2)	99	95	99	100	94	95	94	95	94	94	95	96
	(3)	100	96	100	100	96	96	95	96	96	96	97	98
3	(1)	100	99	101	100	98	98	96	98	98	97	100	100
	(2)	100	96	99	99	95	95	94	95	95	95	96	98
	(3)	100	98	100	100	97	97	95	97	97	96	98	99
4	(1)	101	98	101	101	98	97	96	98	98	98	99	99
	(2)	99	96	99	99	95	95	93	95	96	95	96	97
	(3)	100	97	100	100	97	96	95	97	97	97	98	98
5	(1)	101	98	100	100	98	98	97	98	99	95	100	100
	(2)	99	96	99	99	95	96	95	97	96	96	97	98
	(3)	100	97	100	100	97	97	96	98	98	98	99	99
6	(1)	101	98	100	100	97	98	..	99	98	99	99	100
	(2)	100	96	99	99	95	96	..	97	97	97	97	97
	(3)	101	97	100	100	96	97	..	98	98	98	98	99
7	(1)	101	99	100	100	97	98	98	99	99	99	99	100
	(2)	100	96	99	100	95	95	96	96	96	97	98	97
	(3)	101	98	100	100	96	97	97	98	98	98	99	99
8	(1)	101	99	101	101	98	98	98	98	99	99	100	100
	(2)	100	97	100	100	96	96	97	97	98	97	98	98
	(3)	101	98	101	101	97	97	98	98	99	98	99	99
9	(1)	101	99	100	100	97	98	98	99	99	100	100	100
	(2)	100	98	100	100	96	97	98	98	98	98	98	99
	(3)	101	99	100	100	97	98	98	99	99	99	99	100

* (Maximum (1), Minimum (2) and Mean (3)).

None of the blood specimens at any time showed stippling or other morphology suggestive of degenerative changes of the red corpuscles.

The *bowel movements* during Periods 7 and 8 were on the whole softer in two subjects of Group 1, in one subject of Group 2, and in one subject of Group 3. Otherwise, there were no differences attributable to the dosage. With the two subjects of Group 1, this was accompanied by slight cramps, not sufficiently marked to be reported, but learned only from inquiry.

Aside from slight colds, etc., not related to the dosage, the *general conditions* were negative.

Two other subjects, not reported in detail, ingested during the autumn of 1912 doses equivalent to 2 gm. s.a.s. daily, with negative effects except for slight reduction of urinary phosphates. When mixed s.a.s. and calcium acid phosphates, corresponding to the mixed type of baking were taken, the urinary phosphates were not diminished.

DIGESTION EXPERIMENTS

Biscuits were made with the three types of baking powders, with the ingredients in the proportions:

97.1   S.A.S.   to 100 soda
223.8   Cream of tartar   to 100 soda
115.7   Calcium acid phosphate to 100 soda

In each case the active constituents were diluted with dry starch, so as to make the proportion of the latter 25 per cent.

These were used in the amounts 10 gm., 16.4 gm. and 10.9 gm. respectively to a pint of flour. Salt and lard were added in the same amount for each kind of biscuit and the products baked side by side. After baking, they were air dried for twenty-four hours, and ground to a uniform coarse powder. One-twentieth of each total batch, corresponding to one biscuit, was added to corresponding flasks containing 400 cc. of 0.1 per cent HCl and 0.1 gm. commercial pepsin. This was incubated at 40°C. for two hours with shaking every ten minutes, 0.922 gm. sodium bicarbonate added to each flask and the whole boiled to make a uniform paste of the starch. It was then cooled to 40°C., 0.1 gm. commercial pancreatin with amylolytic and some tryptic activity added and again incubated for five hours. The mixture was brought to a volume of 500 cc. and homogeneous aliquots of 50 cc. used for the tests. After filtration, the nitrogen was determined in the residue and nitrogen and reducing sugars in the filtrate. Corrections were made for nitrogen in the pepsin and pancreatin. The means of 9 tests in each case were:

TABLE LXIII

DIGESTION, SERIES I

	MALTOSE Gm. per Biscuit	NITROGEN Per Cent Insoluble
Phosphate	7.86	16.96
S.A.S.	7.89	19.68
Cream of tartar	7.88	17.21

In explanation of the slightly high insoluble residue of the S.A.S. biscuit, the possibility of the slight solubility of alumina or the harder crust noted with these biscuits is mentioned. In a second series, where the flour weights were less uniform, the influence of the total protein present is indicated.

TABLE LXIV

DIGESTION, SERIES 2

	Av. Wt. gm. Biscuit	N per Biscuit	Per Cent N in Residue
Phosphate	19.1	.2897	13.8
S.A.S.	21.6	.3154	20.7
Cream of tartar	22.4	.3270	21.0

In this series, the crust of the cream of tartar biscuit seemed a little deeper brown, although not overbaked.

It was observed that the solubility of an alum-phosphate residue was less than that of a straight alum. Also, when 0.2 per cent HCl was used, slightly more nitrogen was found in the residues, but better carbohydrate digestion was secured, probably dependent upon the action of the soluble alumina in promoting diastatic action. It is further emphasized that a stronger tryptic ferment would doubtless in all cases have shown a lesser amount of nitrogen in the residue.

"In view of all the experiments, it cannot be said that the aluminum residues in the biscuits exert any action which seriously interferes with digestion. In any event possible interference does not seem to be greater than with other types of powders used for similar purposes."

In regard to the use of alum in food other than baking powders, such as cucumber pickles and certain fruits (maraschino cherries), it is concluded that the residue probably would be no more active than in the residue from baking powder and in consideration of the small amount must be looked upon as unobjectionable.

LONG'S CONCLUSIONS

"It is evident from the investigations recorded that these residues, ingested in relatively large amounts, possess some rather marked properties which must be considered objectionable, and especially as concerns their action on the bowels. While it seems true that this action is due to the saline sulphate rather than to the alumina fraction of the residue, it is still a part of the behavior of 'alum' as practically used. The cathartic action of large residues is objectionable, not only because of the feeling of discomfort sometimes occasioned, but also because of the somewhat diminished nitrogen utilization in consequence of the too frequent stools. A daily ingestion of the residues from 4 gm. of the alum and soda combination must doubtless be considered as objectionable from these points of view, even if no further abnormal conditions result in the person consuming the product.

"But the practical daily consumption is much below the weight here in question, and if we consider only the amount of alum which is important from the standpoint of actual daily consumption the situation is distinctly different. Amounts of alum not above 2 gm. a day do not appear to be harmful in any practical sense. It is true that with the subjects consuming this weight of alum daily there was noticed in the first series of experiments some diminution of the urinary phosphate. But the extent of this diminution was not great enough to be interpreted as an objectionable change in the elimination. The same may be said of the trifling alterations in the acidity which usually follow changes in the phosphate content of the urine.

"It must be kept in mind that in all the applications of alum in food the reaction is such as to decompose the original compounds and leave the phosphate, hydroxide or both, products which are comparatively inert, as is well shown by the results of the digestion experiments.

"With this limitation as to the character of aluminum compounds employed, and it is a practical and legitimate limitation, it cannot be said that, when mixed with foods in the small quantities actually considered necessary, they add a poisonous or deleterious substance or injuriously affect the quality of the food with which they are used."

III. Studies on the Influence of Aluminum Compounds on the Health and Nutrition of Man by Alonzo E. Taylor

A review of the physiological relations of aluminum suggests that it might be expected to act like calcium and magnesium on the one hand and iron on the other, not like the heavy metals (copper and lead) nor like the noble metals. Many of the reactions of aluminum resemble those of calcium, magnesium and iron, particularly the relations of solubility and precipitation. The well-known physiological action of alum when applied to a mucous membrane suggests that if it does possess a demonstrable toxicological action we might expect this to be manifested upon the lining membrane of the alimentary tract. If administration were followed by resorption (which was *a priori* unlikely) we might then expect an action to be displayed upon the liver and kidneys, if these should be the organs of elimination. And, lastly, were aluminum resorbed within the body to form insoluble compounds, these would disturb the normal equilibrium between colloid and salt in protoplasm and cell membrane, and such an action would bring aluminum into the group of so-called protoplasmic poisons.

A further line of investigation more specifically concerns the relations of aluminum to the secretion of hydrochloric acid and to the elimination of phosphates. Aluminum hydroxide is to a certain extent soluble in gastric juice, and within certain limits, this solubility is the function of the concentration of hydrochloric acid. It is possible, therefore, that the presence of aluminum hydroxide in the stomach might lead to alteration or abnormality in the secretion of hydrochloric acid. Within the intestine the aluminum is precipitated in an insoluble state as the hydroxide or as phosphate. There is normally elimination of phosphate both in the feces and in the urine. The factors that determine the route of elimination are not well understood. It is, however, probable that constipation tends to favor greater fecal elimination at the expense of urinary elimination. It is known that the presence of phosphate-binding substances such as calcium increase the output of phosphate in the stool. Like calcium, aluminum deflects phosphoric acid from urinary to fecal elimination, forming an insoluble aluminum phosphate and thus withdrawing a certain quantity of soluble phosphate from resorption. In connection with this reaction two questions are at once suggested: (1) Assuming that increase in the phosphate content of the stools is simply an expression of the combination of aluminum with phosphoric acid within the lumen of the alimentary tract, is this possibly of influence upon elimination or of disadvantage to the elimination of other end-products? (2) Is it possible that aluminum, in addition to combining with preformed phosphate in the lumen of the alimentary tract, is resorbed, and within the body fluids and tissues combines with phosphoric acid, then as aluminum phosphate is eliminated in the intestinal secretions? If the aluminum combines only with preformed soluble phosphate in the contents of the alimentary tract, then the administration of aluminum, even in huge doses, would be comparable to the effect upon the urine of a diet less rich or even poor in phosphates. But if the aluminum is able to enter into the body fluids and tissues and there seize upon phosphoric acid, it is obvious that the phenomenon would have an entirely different and possibly an ominous meaning. The attempt has been made throughout the tests with human subjects, and later in special investigations, to elucidate these relations conclusively; the results, however, were not as definite and clear-cut as is to be desired.

EXPERIMENTS WITH THE FIRST GROUP

Experiments were conducted with a squad of 8 men, 2 control and 6 experimental. The powder was not used in bread, but was administered in wafers or dissolved in water. Six of the subjects took the aluminum compounds, while the other 2 took milk sugar, the men themselves not knowing which they were taking. There were 2 groups of experiments in which the whole squad took part. In the experiments of the first group, which ran from October 8 to December 16, tests were made with potassium aluminum sulphate and s.a.s.

alone. The dose at first was such as to give each man 0.1 gm. (approximately equivalent to a level teaspoonful of alum baking powder) of aluminum a day and was increased from time to time until the daily dose was 0.298 gm. (approximately equivalent to 3 level teaspoonfuls of alum baking powder) of aluminum for each man according to the following schedule of days:

TABLE LXV

SCHEDULE OF DAYS

Subject	1	2	3	4	5	6	7	8
Control.................	70	14	14	14	14	14	14	42
P.A.S. 0.1 gm. Al.......	..	14	14	14	14	14	14	
S.A.S. 0.153 to 0.149 gm. Al.......	..	7	7	7	7	7	7	
0.209 gm. Al.......	..	3*	7	7	7	7	7	
0.298 gm. Al.......	..	14	14	14	..	14	14	14
Control.................	..	14	14	14	28	14	14	14

*Out of squad four days.

The general *diet* list was selected in a consultation with the men, all of whom were students of medicine in the Junior class of the University of Pennsylvania, carrying the same schedule of work and leading the same life. Of the articles of diet selected, each subject fixed the amounts of the various foodstuffs in accordance with his tastes and habits. There was a fixed diet for each day, and the squad was given two weeks in which to fix upon the amounts of foodstuffs required. Once selected, the diet was rigidly maintained; alcohol, irregular hours, irregular exercise, and all violent exercise were interdicted. The *total ingested matter* varied from 2190 gm. to 3700 gm. daily, with average values ranging from 2500 gm. to 3200 gm. for the different subjects.

The general *clinical observations*, controlled by the physician of the squad, Dr. Alfred Stengel, but with the intelligent cooperation of students of medicine, included morning and evening temperature and pulse, state of appetite and sensation of health; with notations as to malaise, headache, nausea, colic and number and character of the stools, all of which were normal, excepting as described. With increasing amounts of s.a.s. ingested, symptoms of gastrointestinal disturbance appeared, consisting chiefly of colic (pains not severe), a tendency to constipation with occasionally a soft stool, and in some instances loss of appetite. Symptoms were more marked in some subjects than in others. Subject 1 (control) recorded numerous references to gastrointestinal disturbances while Subject 4 (experimental) presented no such symptoms, yet it seems clear that colic is to be attributed to the large doses of s.a.s. administered. We may therefore summarize the data to the effect that large doses of sodium aluminum sulphate tend to produce a somewhat characteristic type of gastrointestinal irritation with colic. It must be pointed out, however, that the doses termed large were in reality very large. Occasional headaches and colds noted were not related to the aluminum administrations.

The *blood* hemoglobin and morphology examined at weekly intervals presented only the normal variations without relation to the aluminum ingestion.

The following hemoglobin percentages were reported:

TABLE LXVI

HEMOGLOBIN (PER CENT)

Subject	1	2	3	4	5	6
Date						
October 9	85	80	83	84		
10	..	..	..	..	85	82
16	85	84	87	86		
17	..	..	..	..	83	78
23	90	86	87	88		
24	..	..	..	..	86	85
30	90	88	90	88		
31	..	..	..	..	88	80
November 6	88	90	89	93		
7	..	..	..	..	88	80
13	92	89	90	94		
14	..	..	..	..	85	80
20	92	93	94	96		
21	..	..	..	..	88	82
27	92	95	89	97		
28	..	..	..	..	90	82
December 4	90	92	88	86		
5	..	..	..	..	90	76
11	90	88	93	92	93	85

By reference to the previous table it will be seen that Subject 1 was a control throughout and that with the remaining subjects, prior to October 22 and subsequent to December 2, aluminum was not administered (after Nov. 18 with Subject 5).

The *urine* presented only the usual variations in volume and specific gravity; in only one subject it contained a trace of albumin and that twice in the control and once in the experimental periods; it contained casts during the experimental periods only when similarly present in the beginning control period; it was similarly negative regarding other elements in the sediment, and negative for sugar throughout. Aluminum was not present in the urine.

The *feces*, comparing the control and maximum aluminum administration periods, in 6 experimental subjects showed some increase in daily weight (av. 111.1 gm. to 143.7 gm.) due both to increase in moisture (av. 76 to 79.1 per cent) and solids (av. 25.8 gm. to 28.7 gm.), including to a slight extent an increase in fatty acids (av. 1.8 gm. to 2.2 gm.) but not in proteins. In a single experimental subject, this order was reversed (weight 167.5 gm. to 105.6 gm.; moisture 81.9 to 74.8 per cent; solids 30.3 gm. to 26.8 gm.; fatty acids 2.2 gm. to 1.8 gm.).

The occurrence of colic ran parallel to the increase in the weight of the fresh stools and in the increase in water content.

The *nitrogen* input ranged from 13 gm. to 16 gm. per day while the output of all subjects in the experimental periods displayed no variations between the control and experimental periods. Likewise, nitrogen utilization was not interfered with during the aluminum periods, despite the increase of the feces and of the moisture content.

The *phosphorus* input was quite constant as also the total output. The balance showed slight retention. With increasing administration of s.a.s. there was a corresponding diminution of urinary and increase of fecal phosphorus, a deflection attributed to the aluminum compound. This varied from 10 to 25 per cent of the urinary phosphorus.

TABLE LXVII

PHOSPHORUS INGESTION AND ELIMINATION

(Daily Average Grams*)

Subjects		1	2	3	4	5	6	7	8
Days									
1–14	Added Al	0	0	0	0	0	0	0	0
	P. food	1.53	1.55	1.48	1.46	1.62	1.38	1.55	1.68
	urine	1.04	0.96	0.94	1.01	1.01	0.99	1.05	1.12
	feces	0.33	0.49	0.36	0.40	0.45	0.31	0.44	0.41
	Balance	+.16	+.10	+.18	+0.05	+0.16	+0.08	+0.06	+0.15
15–28	Al, p.a.s.	0	0.100	0.100	0.100	0.100	0.100	0.100	0
	P. food	1.50	1.53	1.46	1.45	1.58	1.37	1.51	1.67
	urine	1.08	0.93	0.86	0.99	1.01	0.90	1.00	1.11+
	feces	0.32	0.49	0.44	0.42	0.45	0.44	0.45	0.41+
	Balance	+.10	+.09	+.16	+0.04	+0.12	+0.03	+0.06	+0.15
29–42	Al, s.a.s.‡	0	0.153 to 0.209†	0.153 to 0.209	0.153 to 0.209	0.153 to 0.209	0.153 to 0.209	0.153 to 0.209	0
	P. food	1.48	1.45	1.46	1.45	1.59	1.36	1.53	1.66
	urine	1.01	0.82	0.86	0.90	0.91	0.84	0.93	1.06
	feces	0.33	0.58	0.44	0.44	0.51	0.41	0.57	0.38
	Balance	+.14	+.05	+.16	+0.11	+0.17	+0.11	+0.03	+0.22
43–56	Al, s.a.s.	0	0.298	0.298	0.298	None	0.298	0.298	0.298
	P. food	1.48	1.50	1.44	1.44	1.56	1.35	1.52	1.65
	urine	1.02	0.77	0.85	0.91	1.02	0.80	0.86	1.00
	feces	0.40	0.65	0.50	0.53	0.42	0.44	0.61	0.56
	Balance	+.06	+0.08	0.09	0.00	+0.12	+0.11	+0.05	+0.09
57–70	Al	0	0	0	0	0	0	0	0
	P. food	1.48	1.53	1.45	1.46	1.59	1.35	1.54	1.66
	urine	1.03	0.87	0.92	0.99	0.95	0.90	0.99	1.10
	feces	0.39	0.43	0.38	0.37	0.43	0.35	0.55	0.43
	Balance	+0.06	+0.23	+0.15	+0.10	+0.21	+0.10	0.00	+0.13

* From Daily Balance Charts.

† No Al last 4 days.

‡ Actual daily amounts .153, .153, .149, .149, .149, .149 .149, .209, .209, .209, .209, .209, .209, .209.

The *urinary nitrogen partitions* showed only the usual variations of normal individuals.

Aluminum recovered in the stools amounted to 97, 99, 99, 100, 100+, 101.3 and 101.7 (av. 99.7) per cent of the amount ingested in the corresponding experimental subjects. The figures are regarded as a quantitative output, that is, not retained in the body but completely eliminated; but this is not regarded as proof that the aluminum did not leave the alimentary tract. Further data on this point were obtained in a special research.

EXPERIMENTS WITH THE SECOND GROUP

In the experiments which ran from January 14 to May 10, 1912, tests were made with various aluminum compounds or sodium sulphate, simulating more or less closely the residues that might be expected to form in alum baking powders, as per the schedule of days shown in Table LXVIII.

The *total ingested* matter of the diet varied from 2160 gm. to 3540 gm. daily, with average values varying from 2700 gm. to 3220 gm. daily for the different subjects.

General *clinical observations:*

Subject 1 showed occasional headaches and malaise with, at times, blood in stools from anal fissure; slight colic during the administration of sodium sulphate; appetite usually good or fair. He was chiefly a control subject. There was no relation of symptoms to aluminum administration. Body weight remained constant.

Subject 2 showed no symptoms during the first twenty-eight days. During the next week he experienced poor appetite, some headache, nausea and colic with soft stools; during the sodium sulphate period, colic with soft and watery stools; day 67, slight headache and malaise; 80 to 83, headache, nausea, tongue coated, colic; 91, colic. Body weight showed gain of 1½ lb.

Subject 3 showed: days 4 to 6, coryza, watery stool; 30 to 54, stools soft, occasionally watery, and slight colic; 58 to 60, coryza, 65 to 88, stools normal or hard with constipation; 88, headache and nausea; 89 to 95, sick in hospital; 101, slight subicteroid persisting to close of experiment. Body weight showed gain of 2 lb.

Subject 4 showed: days 10 to 14, sore throat, followed for 2 weeks by nasal catarrh; 25, 26, 31, 32, poor appetite, soft and watery stools; 52 to 55, headache, gurgling, watery stools; 66 to 68, sore throat; 95 to 99, poor appetite, distention, belching, tongue coated. Body weight was constant.

Subject 5 showed: days 6, 22 to 24, 46, 49, 52 to 54, 57, 63, 101, 103, 108, slight headache; 10, 11, 16, 18, 41, 72, 76, 100 to 101, 104, slight abdominal pain and soreness; 21, 64 to 77, cold and sore throat; 32, 39, 45 to 46, 50 to 54, 57 to 59, 80 to 81, 84, 90, 92, 95 to 97, 104, 106 to 107, 116,

TABLE LXVIII

SCHEDULE OF DAYS*

Subjects	Al mgs.	1	2	3	4	5	6	7	8
Periods...									
Fore...	...	1 to 14‡	1 to 14 15 to 28 29 to 35	1 to 14 15 to 28 29 to 49 50 to 53 54	1 to 14 15 to 28 29 to 49 50 to 53 54	1 to 14 15 to 28 29 to 49 50 to 51 (52 to 56)	1 to 14 15 to 28 29 to 49 50 to 51 52 to 56‖	1 to 14	1 to 14 15 to 28 29 to 35
S.A.S } NaHCO₃ } Residue§	227 498 747 498								
Control. Na₂SO₄		15 to 49	(36 to 42) 43 to 49	55 to 63	55 to 63	57 to 63	57 to 63	15 to 42 43 to 49	36 to 42 43 to 56 57 to 63
5.23 gm. Na₂SO₄		50 to 56	50 to 56						
Control.		57 to 63	57 to 63					50 to 77	
Al(OH)₃ Merck	476	64 to 70	64 to 77	64 to 77	64 to 77	(64 to 77)	64 to 77		64 to 77
Al(OH)₃ Emulsion	558	(71 to 84)	78 to 84	78 to 84	78 to 84	78 to 91	78 to 84	78 to 84	78 to 84
Control.	249 498	85 to 105	85 to 105			92 to 97 98		85 to 105	85 to 91 92 to 98
S.A.S } NaHCO₃ } Residue§	485			85 to 105¶	85 to 91 92 to 98 99 to 105	99 to 105	85 to 105		99 to 105
AlCl₃	727								
NaHCO₃ } Residue	970								
After...		106 to 118	106 to 118	106 to 118	106 to 118	106 to 118	106 to 118	106 to 118	106 to 118

* Figures are the days of observation; figures in parentheses are days off squad.

† Administered daily.

‡ Day 12, some bleeding from hemorrhoids.

§ Commercial aluminum baking powder, dissolved in water, evaporated to dryness and baked.

‖ Day 54, no aluminum residue administered.

¶ Days 89 to 95, off squad.

colic, usually slight; 42, 51 to 56, 103, slight nausea; 51 to 54, 52, vomiting; 111 to 112, poor appetite; 52, 102, subicteroid; 107 to 118, gastrointestinal attack. Body weight remained constant.

Subject 6 showed: days 6, 28, 30, 47, 52, 78, 87 to 88, 118, headache; 6, 47, 118, cold; 28, 30, toothache; 36 to 37, 45 to 46, 49, 51, 53, 55, 89, 112, 117, diarrhea or watery stools; 52 to 53, 93, colic; 58, 99, 101, epigastric pain; 60, feeling of faintness; 62, nausea; 85, dizziness; 87 to 88, earache; 29 to 30, 62, poor appetite; 52 to 118, tongue coated. Body weight showed loss of 2 lb.

Subject 7 showed: days 24, headache; 43, slight nausea; 46 to 52, 60, 112, colic; 46 to 47, watery stools; 46, 102, subicteroid. Body weight showed gain, 3½ lb.

Subject 8 showed: days 30 to 31, 37, 52, 87, 100, headache; 34, slight cold; 31, 35, 37, 85, 97 to 101, colic; 98 to 102, nausea; 26 to 35, moderate looseness; 102, epigastrium distended, pain and tenderness over lower bowel. Body weight showed loss of 2½ lb.

The physician of the squad explained the colic and diarrhea by the administration of varying amounts of alkaline sulphates. Some of the interruptions were purely accidental, being due to attacks of tonsillitis, bronchitis, or the like. Clinically, there was no evidence to show injury other than from repeated administration of a laxative.

The author emphasizes that the most noteworthy symptoms are to be ascribed to the sodium sulphate and that in the absence of this, colic and abdominal distress of less definite type follow the ingestion of aluminum hydroxide in very large amounts.

As with Squad 1, the *urine* was negative throughout, as to evidence of renal irritation and sugar. Subject 3 showed a trace of albumin on the twentieth day and Subject 4 on the seventy-fourth day. Casts were rarely found and then as frequently as during the control periods.

The *feces* showed with the ingestion of large amounts of residue of aluminum baking powder an increase in total weight and water contents, similar to the increase noted when corresponding amounts of sodium sulphate were alone administered. A slight increase of fat was noted in some instances.

The *blood*, as to hemoglobin and morphology, was entirely negative throughout.

The following hemoglobin percentages were reported:

TABLE LXIX

HEMOGLOBIN (PER CENT)

Subject	1	2	3	4	5	6	7	8
Day								
2	80	90	88	90	84			
3	..	..	..	..	..	75	80	82
9	..	90	88	90				
10	75	..	..	..	84	65	72	85
16	79	89	87	90				
17	..	..	..	..	85	80	84	82
23	78	90	88	90				
24	..	..	..	..	84	69	80	86
30	68	90	85	90				
31	..	..	..	..	82	68	84	78
37	72	..	84	88				
38	..	..	..	..	85	70	83	83
44	70	78	84	86	..	..	84	
45	..	..	..	..	82	75	..	84
51	74	83	84	89				
52	..	..	..	..	88	75	86	84
58	79	82	85	83				
59	..	..	..	..	84	78	84	84
65	78	83	80	88				
66	..	..	..	..	..	76	85	72
72	..	88	85	87				
73	..	..	..	..	..	78	84	84
79	..	84	88	86				
80	..	..	..	..	84	76	82	83
86	80	84	84	88				
87	..	..	..	..	85	79	86	83
93	82	83	..	80				
94	..	..	..	..	83	75	86	80
100	79	85	84	87				
101	..	..	..	..	83	80	86	87
107	83	86	86	88				
108	..	..	..	..	86	78	83	80
114	80	87	85	89				
115	..	..	..	..	78	78	83	87
131	82	88	89	92				
132	..	..	..	..	88	80	86	85

Average of Subjects 2 to 8 on Days 2, 3, 84½ per cent; average of Subjects 2 to 8 on days 100, 101, 84½ per cent.

Fluctuations were not different or greater in the experimental subjects than in the control, or in the experimental periods than in the control periods of the same subjects.

The average *alimentary utilization* of nitrogen was slightly lower in 4 of the subjects during the administration of sodium sulphate with or without the administration of aluminum, was not lower when the aluminum was administered with sodium sulphate, and is ascribed to the laxative action of the sodium sulphate.

The *urinary nitrogen partition* showed only the usual variations of normal individuals.

The *distribution of the nitrogen* output between the urine and feces was normal throughout and closely constant.

With the aluminum baking powder and hydrate intake there was as a rule a proportional deflection of the urinary to the fecal *phosphorus;* also to a less extent with aluminum chloride. The phosphorus of the stools might rise 20 per cent, and of the urine, fall 10 to 15 per cent. In this series, it was less marked than in the experiments with Squad 1 and showed some striking exceptions.

Aluminum recovered in the stools amounted to 99, 98, 98, 97.5, 95.4, 97.7, 99.6 and 97.8 (av. 97.5) per cent of the amounts ingested in the corresponding experimental subjects.

There were no nitrogen deficits and practically no phosphorus deficits in the *metabolic balances* and it is concluded that the aluminum administration did not in any discernible way disturb the normal flow of the total metabolism of these elements.

The phosphorus balance remained practically constant, not being appreciably influenced by the ingestion of large quantities of aluminum residue, as the following data show:

TABLE LXX

PHOSPHORUS BALANCE

DAILY AVERAGES

Subject 1

Days	Al, mg. administered	Food, gm.	Urine, gm.	Feces, gm.	Balance, gm.
1 to 14	....	1.33	0.90	0.39	+.04
15 to 28	....	1.32	0.84	0.41	+.07
29 to 49	....	1.35	0.90	0.40	+.05
50 to 56	(5.23 gm. Na₂SO₄)	1.36	0.89	0.48	−.01
57 to 63	....	1.38	0.92	0.42	+.04
64 to 70	476	1.37 .	0.87	0.48	+.02
85 to 105	....	1.36	0.87	0.42	+.07
106 to 118	....	1.35	0.88	0.53	−.06

Subject 2

Days	Al, mg. administered	Food, gm.	Urine, gm.	Feces, gm.	Balance, gm.
1 to 14	....	1.35	0.79	0.43	+.13
15 to 28	227	1.33	0.78	0.45	+.10
29 to 35	498	1.35	0.69	0.57	+.09
43 to 49	....	1.36	0.81	0.48	+.07
50 to 56	(5.23 gm. Na₂SO₄)	1.41	0.88	0.49	+.04
57 to 63	....	1.43	0.84	0.49	+.10
64 to 77	476	1.43	0.80	0.45	+.18
78 to 105	558 } 485 }	1.48	0.76	0.53	+.19
106 to 118	....	1.45	0.85	0.42	+.18

Subject 3

Days	Al, mg. administered	Food, gm.	Urine, gm.	Feces, gm.	Balance, gm.
1 to 14	....	1.19	0.78	0.34	+.07
15 to 28	227	1.17	0.70	0.41	+.06
29 to 56	536	1.20	0.66	0.53	+.01
57 to 63	....	1.22	0.78	0.38	+.06
64 to 77	476	1.21	0.76	0.38	+.07
78 to 88	531	1.21	0.59	0.66	−.04
96 to 105	485	1.21	0.77	0.49 .	−.05
106 to 118	....	1.19	0.73	0.36	+.10

Subject 4

Days	Al, mg. administered	Food, gm.	Urine, gm.	Feces, gm.	Balance, gm.
1 to 14	....	1.19	0.85	0.31	+.03
15 to 28	227	1.17	0.76	0.39	+.02
29 to 56	536	1.20	0.71	0.48	+.01
57 to 63	....	1.22	0.82	0.29	+.11
64 to 77	476	1.20	0.82	0.37	+.01
78 to 91	522	1.19	0.68	0.49	+.02
92 to 105	848	1.22	0.77	0.43	+.02
106 to 118	....	1.19	0.86	0.31	+.02

Subject 5

Days	Al, mg. administered	Food, gm.	Urine, gm.	Feces, gm.	Balance, gm.
1 to 14	....	1.42	0.93	0.38	+.11
15 to 28	227	1.40	0.90	0.37	+.13
29 to 52	520	1.41	0.87	0.50	+.04
57 to 63	....	1.45	0.94	0.43	+.08
78 to 91	....	1.46	1.00	0.39	+.07
92 to 105	385	1.47	0.99	0.43	+.05
106 to 118	....	1.37	0.96	0.38	+.03

TABLE LXX. (*Continued*)

Days	Al, mg. administered	Food, gm.	Urine, gm.	Feces, gm.	Balance, gm.
		Subject 6			
1 to 14	....	1.39	0.80	0.59	0.0
15 to 28	227	1.36	0.79	0.50	+.07
29 to 56	498	1.40	0.82	0.59	−.01
57 to 63	....	1.41	0.80	0.52	+.09
64 to 77	476	1.39	0.78	0.58	+.03
78 to 105	503	1.40	0.74	0.57	+.09
106 to 118	....	1.37	0.89	0.42	+.06
		Subject 7			
1 to 14	....	1.32	0.86	0.43	+.03
15 to 42	....	1.32	0.81	0.46	+.05
43 to 49	(5.23 gm. Na2SO4)	1.33	0.89	0.42	+.02
50 to 77	....	1.34	0.83	0.47	+.04
78 to 105	503	1.33	0.73	0.55	+.05
106 to 118	....	1.31	0.79	0.44	+.08
		Subject 8			
1 to 14	....	1.37	0.96	0.41	0.0
15 to 28	227	1.36	0.86	0.44	+.06
29 to 35	498	1.39	0.88	0.51	0.0
36 to 42	....	1.37	0.94	0.36	+.07
43 to 56	(5.23 gm. Na2SO4)	1.37	0.96	0.39	+.02
57 to 63	....	1.39	0.98	0.43	−.02
64 to 77	476	1.38	0.91	0.43	+.04
78 to 91	522	1.38	0.84	0.58	+.04
92 to 105	848	1.40	0.87	0.45	+.08
106 to 118	....	1.36	0.93	0.38	+.05

ON THE CHEMICAL RELATIONS OF ALUMINUM IN THE ALIMENTARY TRACT AND THE CONDITIONS OF ITS POSSIBLE ABSORPTION

Aluminum hydroxide is to a certain extent soluble in dilute hydrochloric acid and aluminum phosphate to a very slight extent; they may, however, form colloid precipitations with proteins that are not soluble in dilute hydrochloric acid. Such solutions lack the direct local action which a piece of sodium aluminum sulphate exhibits when placed upon an abrasion of the skin or mucous membrane. In the intestine, on neutralization of the hydrochloric acid, the aluminum is precipitated as hydrate and phosphate.

In the two periods of experimentation, 270.55 gm. of aluminum were administered and 265.71 gm., or 98.2 per cent, recovered in the stools. In the analysis of feces there is nearly always a loss, so that the recovered aluminum may be taken to approximate the input and Taylor believed there had been no retention of aluminum. If resorbed it must have been promptly eliminated by the mucous membrane of the intestine, and in the bile. He had never been able to demonstrate the presence of aluminum in the urine.

Steel reported the presence of a few milligrams in the blood of the dog, following the administration of large doses of aluminum. "We* have made the test upon the human being." Four subjects were given for a period of several days ingestions of aluminum amounting to about 1 gm. of the metal per day. From a vein in the arm of each man 200 cc. of blood were drawn. In this amount (over 800 cc.) it was not possible to demonstrate the presence of aluminum. All reagents and apparatus (quartz and platinum) were demonstrated free from aluminum.

Experiments were conducted to determine whether an increase of phosphate elimination, when the diet was poor in phosphate, demonstrated abstraction of phosphate from the body fluids and tissues such as might occur if aluminum were resorbed in a form other than phosphate, and eliminated as phosphate.

Experiment 1. Phosphorus intake (daily) 0.37 gm. Fore period, five day, phosphorus output, urine 0.55 gm., feces 0.12 gm., total 0.67 gm., loss of 0.3 gm. Exp. period, five day daily ingestion of Al (hydrate) 0.660 gm. Phosphorus output, urine 0.30 gm., feces 0.36 gm., total 0.66 gm., loss of 0.29 gm.

Experiment 2. Phosphorus intake (daily) 0.17 gm. Fore period, output phosphorus, urine 0.53 gm., feces 0.10 gm., total 0.63 gm., loss of 0.46 gm. Exp. period, five day, daily ingested Al (hydrate) 0.540 gm. Output phosphorus, urine 0.29 gm., feces 0.28 gm., total 0.57 gm., loss 0.40 gm. After period, output phosphorus urine 0.42 gm., feces 0.10 gm., total 0.52 gm., loss 0.35 gm.

The figures fail to substantiate the assumption that aluminum was resorbed and robbed the body of phosphorus, and hence fail to give evidence of the resorption of aluminum. "Resorption of traces we have obviously not disproved."

*Taylor.

Deflection of the elimination of phosphorus from the urine to the feces tends to reduce the acidity of the urine, but not more than is induced by variations in diet or by the ingestion of calcium such as occurs, for example, when hard water is drunk.

Dr. Alfred Stengel, who made the clinical examinations of all subjects in all the foregoing tests stated: "So far as my clinical observations could be relied upon, there was no evidence to show any injury other than that to be expected from such repeated administration of a laxative."

TAYLOR'S CONCLUSIONS

"It is difficult to summarize the experiences and difficult also to evaluate the results in an absolute manner. We have had unquestionably evidences of the catharsis caused by the administration of large doses of baking powder. Such ingestions of baking powder as occur in ordinary life were not tested in our experiments. The lowest dose employed by us was higher than the ingestion one would meet with in the ordinary use of cake, bread and biscuits prepared with baking powder. Campers consume in a day as much as our medium dose and occasionally campers consume as much as our largest regular dose. It is obvious, therefore, that the conditions of the experiments as carried through by us represent in no wise the usual, but rather the very unusual, use of baking powder. Under these circumstances the stools are loose, colic is apt to attend the evacuations, the actual amount of water in the stools is notably increased, and here and there is to be observed an increase in the solids. A comparison of the results attending the administrations of sodium aluminum sulphate, aluminum hydroxide, aluminum chloride plus sodium bicarbonate, aluminum baking powder residue, and sodium sulphate indicate that catharsis of the types described is to be regarded as the result of sodium sulphate, a residuum of the reaction. With very large doses of sodium aluminum sulphate, aluminum hydrate, and aluminum chloride plus sodium bicarbonate, occasional dry colic may be noted.

"The practical ultimate interpretation of these findings is difficult to formulate. I personally do not believe that it would be healthful for anyone, in camp or out of camp, to live upon a diet of baking powder biscuits. To what extent the one-sided diet of wheat flour, prepared in this form, would exaggerate the effects of the saline catharsis, cannot be stated. I do not believe that the regular ingestion of sodium sulphate in doses of from 3.5 gm. to 5 gm. per day, with the normal diet, resulting in distinct looseness of the bowels, is a procedure to be recommended. Prolonged administration of saline cathartics, even in small doses, tends to leave behind a condition of constipation; and it is certainly the experience of the medical profession that the practice of the regular administration of saline cathartics is not to be recommended. This aspect of the question is of course not peculiar to aluminum baking powder, but applies to all baking powders, since to a greater or less extent a saline cathartic remains as the residue of the reactions of all known baking powders. There is no evidence in our results to indicate that the occasional and ordinary use of bread, biscuits, or cake prepared with aluminum baking powder tends to injure the digestion. The amount of saline cathartic that would be ingested, under conditions of normal diet, would be very small and would provoke no catharsis or symptoms of any kind. There is some question of the desirability of the use of cake, bread and biscuits in conditions of indigestion; but condemnation is not based upon the fact that they are prepared with baking powder, but rather upon their physical properties and the fact that these articles are usually consumed either hot or within a few hours after baking.

"The possibility of injury involved in the use of very large ingestions of aluminum, such as have been shown to provoke occasional dry colic, is difficult of determination. If aluminum were ingested only in the state of aluminum baking powder (as is practically the case, since the possible amount in other foods is minimal), it is obvious that the high ingestions which in our experiments were large enough to cause colic could never occur outside of camp life."

The problem and conclusions of the Board are presented on page 92.

Abstract of the Testimony of
Russell H. Chittenden[*4]

Among the problems presented to the Board was one under the general caption: Do aluminum compounds when present in food affect injuriously the nutritive value of such foods or render them injurious to the body?

At various conferences in which all the members of the Board were present and participated it was agreed that there should be three independent series of investigations, one by Dr. Long in Chicago, the second by Professor Alonzo E. Taylor, then of the University of Pennsylvania, and the third by Russell H. Chittenden at New Haven.

The whole Board participated in the study of the general problem of how to proceed in the case, the general method to be followed and the general trend to three distinct investigations. . . . Consultations were held and the members visited each other's laboratories during the period of the work. When the work was completed a conference was held to consider all the results in detail. Then each member individually wrote his report and these reports were scrutinized by all the members of the Board. Finally there was a brief, concise report to the Secretary of Agriculture, embodying in a few pages the general conclusions reached by the Board and signed by the five members. This was submitted to the Secretary along with the more detailed reports of the individual workers.

After careful reading at the time it appeared and since, Chittenden believed that Bulletin No. 103,[5] Department of Agriculture, based upon the report submitted by the Referee Board but issued solely and entirely by the Department of Agriculture, embodies faithfully the conclusions arrived at by the Referee Board.

At the outset of the work, all available literature, in English as well as in French and German, was gone through very thoroughly and examined with a view of determining the state of knowledge at that time with regard to the physiological effect of various aluminum compounds.

* Given at the request of the Federal Trade Commission, December 5 and 19, 1924.

Quantities of Aluminum. One of the first problems considered by the Board in outlining its own work was what should be looked upon as a small quantity and what should be looked upon as a large quantity, in view of the specific form of the question presented by the Department of Agriculture.

After a very careful consideration, it was decided that an amount of aluminum in the daily food amounting to from 25 mg. to 75 mg. of aluminum should be considered the small quantity and that the larger quantity should comprise up to 150 mg. or even 200 mg. of aluminum per day. It was considered very important, however, that the Board learn everything possible with regard to the effect of aluminum compounds as used in baking powder and the leavening of bread, and therefore it was decided to experiment with quantities as large as 1000 mg. of aluminum a day.

Scope of Individual Experiments. In view of the decisions, it became desirable to mark out rather definitely the scope of each man's work. It was decided that in the investigation by Chittenden at New Haven stress should be laid upon the effect of the smaller doses of aluminum salts. As a matter of fact, his data showed that for a period of three months, ninety days, the average amount of aluminum taken by each subject was 68 mg., ranging from 65 mg. a day to 75 mg. a day, with an average daily intake of 68 mg. of aluminum in three months. The dose was then increased up to a maximum of 260 mg. per day, that portion of the work covering forty days.

Furthermore, it was decided in the New Haven experiments to introduce the aluminum as baking power in bread, in other words give the aluminum to the subject as bread leavened with alum baking powder. This is the form in which all the subjects at New Haven received the aluminum.

Chittenden had in all 12 subjects, 8 of whom received aluminum for one hundred and thirty days consecutively, and 4 served as controls. The maximum dosage was 260 mg.

In the experiments carried out by Dr. Long and Dr. Taylor, it was decided that stress should be laid upon the effects of larger quantities, and furthermore that the aluminum should be given in a different form. Accordingly, as the report shows, the aluminum was introduced (1) as aluminum hydroxide, or as aluminum chloride treated with the requisite quantity of sodium bicarbonate to give the reaction that takes place in the leavening of bread; and (2) mixed with milk, (3) with water and (4) in wafers or in tablets. In other words, the idea of the Board was to introduce the aluminum in different forms from those in which it would ordinarily enter the system when taken in baking powder bread, in order to bring in variety and give the aluminum an opportunity to exercise any such injurious effects as it might possess. In these two series of experiments, which occupied a shorter time than Chittenden's, the dose was considerably higher. Thus, by varying the form and the quantity of the administrations, the work as a whole covered a wider scope.

There was also a difference in procedure in regard to the analytical work carried on.

In the New Haven experiment emphasis was laid upon the idea of studying the effect upon digestion, absorption and utilization of the food, upon the utilization of the nitrogen or protein, fat utilization, etc., whereas in some of the other experiments more stress was laid upon determining the various ways in which the nitrogen was eliminated from the body. Questions of phosphorus balance, questions with regard to the absorption of aluminum by the blood, and the elimination of aluminum by the urine and by the feces, all were presented to give variety and cover the ground as broadly as possible. In addition, of course, in every case the subjects were under very careful medical observation; at New Haven they were examined by the medical expert every ten days during the whole five months. During all periods there were blood examinations including the hemoglobin content and blood counts, the leucocytes and red corpuscles. The appearance of the blood corpuscles was very carefully studied by Professor Rettger in the bacteriological laboratory, and to a lesser extent by Professor Rand.

DETAILS OF EXPERIMENTS GIVEN IN THE EVIDENCE

"Q. Then to sum up, Dr. Chittenden, it is correct to say that so far as you were able, you took all the known ways of finding out completely the effect of the ingestion of aluminum compounds in the ways you described?

"A. We so considered it.

"Q. Now, will you tell us the results, I mean a summary of results of the work of these investigators?

"A. I think the summation is very clearly expressed in the final report of the Referee Board. . . . Certainly, in quantities up to 150 mg. of aluminum a day there were absolutely no indications of any injurious or abnormal effect whatever. When a dosage of 200 mg. of aluminum was reached there was a tendency toward a mild catharsis. That was worked out very carefully, and it was the opinion of the Board, in which we all concurred, that that was due practically entirely to the sodium sulphate. In my own experiments when a dosage of 260 mg. of aluminum a day was given, the amount of sodium sulphate present with that quantity would be 3½ gm., so that we might expect, especially in sensitive individuals, a slight catharsis. When the dosage went far beyond it, the catharsis was more pronounced, and Dr. Taylor, particularly in the very large doses which he used, got quite marked catharsis.

"In my own experiments, going up as high as we did, there was absolutely no sign of any cathartic action at all. The only thing noticeable was a slight tendency for the weight of the feces to increase and a slight tendency toward a little higher percentage of water in the feces.

"Q. Did the Board make any tests to see whether other forms of baking powder would exercise a cathartic action?

"A. None in my laboratory; but Dr. Long made some experiments, which I did not see, of course, but which I heard about. It seemed to the Board, however, that real experiments of that sort were hardly necessary because of the well-known action of sodium sulphate, sodium phosphate and sodium tartrate, all producing more or less catharsis when given in certain quantities, and perhaps, especially the tartrate, a little diuretic action . . .

"Q. Will you describe in more detail, Dr. Chittenden, your own experiments which were made a part of the Referee Board report?

"A. In the first place I would like to emphasize the fact that the men who composed the squad taking the alum baking powder bread were men ranging in age from nineteen to thirty-four years. They were men not selected for robustness, but they were, of course, chosen with a view to general good health so as to avoid introducing any unnecessary variables into the problem.

"They were fed in the Yale dining hall in a special room under our own supervision, and the bread used as a vehicle for the aluminum was made in our own bakery by a special baker, who was under our own control, with the baking powder which was made in the laboratory of sodium aluminum sulphate, sodium bicarbonate and corn starch. The bread (alum bread) was baked each day with a definite content of aluminum, but in order to recheck the amount of aluminum which the men were taking, samples of the bread were analyzed for aluminum each day.

"The 4 control subjects were fed on bread leavened with yeast which we had made in the bakery of the dining hall.

"The urine of these men was collected every day and subjected to analysis. The feces were collected and weighed each day and subjected to analysis in given periods of ten days, and the feces were separated by means of carmine stain so as to get a sharp division.

"The men were weighed each day. The total nitrogen in the urine was determined and the food was analyzed each day. The food for each man was weighed each day. During the fore-period of four weeks, we found out the natural inclinations of the subjects, so that they were allowed to eat about what they wanted; of course we measured and weighed every particle of food that they consumed and then samples were analyzed.

"It is interesting, however, as an illustration of the variety of the diet, that the average daily intake of nitrogen in the alum subjects ran from 13.9 gm. to 21.6 gm. and the caloric value was about 3000 calories. The phosphate phosphorus was determined in the urine each day so as to get some light on the elimination of phosphorus; and by the results of the analytical data which we obtained we were able to calculate the exact values with regard to the utilization of the protein food and the utilization of the fats, remembering of course that possibly the aluminum salt might exercise some deleterious effect on the digestion or on the utilization of either protein or fatty food. We found, however, no indication of any such result. All through that long period of three months with the smaller doses, and even during the month and a third with the larger doses, there was complete utilization of the protein food and complete normal utilization of the fatty food, as measured by comparison with the control subjects.

"It is understood, I assume, that in a study of this kind, examination of the daily excreta to determine the metabolic action is as sure a way as any of measuring changes that may take place in the tissues or organs of the body. Many disturbances, even lesions, are first suggested to the expert physiologist or physiological chemist by the recognition of some change in the excretions of the body.

"It is a legitimate, safe and reliable method of ascertaining whether normal processes are prevailing or not. In connection with this I am reminded that in 1907 a German by the name of Blum and his associate found that acids having two carboxyl groups, when introduced into the system, brought about a very decided reduction in the excretion of urinary nitrogen; but it was not until 1912 or thereabouts that Professor Underhill in my laboratory, aided later by Professor Wells, found that one such acid and its salts brought about a distinct disintegration of the epithelial cells of the tubulae of the kidney.

"In the New Haven experiments, during all this time, daily examinations of the urine showed no abnormal constituents. There were one or two cases of slight reactions for albumen, but these were as marked in the control subjects as in the alum subjects. Further, in following out this point, there was no relationship between the albumen reaction, the slight traces and the dosage of aluminum compound. Nothing abnormal, nothing irregular, in the composition of the urine was found during the experiment to indicate any deleterious effect of the aluminum compound.

"Examinations of the blood were under the personal supervision of Professor Rettger who made examinations at stated intervals all through the series of experiments. The medical examinations of the men were made at ten day periods by Dr. Rand of the Yale Medical School, who reported on his findings each period and at the end. He said in his final report that he could find no evidence that the feeding in any way affected the general health and nutrition or digestion or had any effect on the excretory functions of the kidneys or intestines. He determined mouth temperature, pulse rate, blood pressure, examined the heart and lungs, spleen and liver, peristaltic disturbances, gas formation in the intestines, etc.

"Q. After the publication of Bulletin 103 by the Department of Agriculture did you see in the scientific literature any reports of subsequent investigations concerning the effect of aluminum in foods which indicated any error in the conclusions of the Board?

"A. I have not seen any results which would lead me to change my conclusions, which are a part of the conclusions of the Referee Board.

"Q. If you were asked the same questions today which were presented to you during the Referee Board work, would you answer them in the same way?

"A. I should, decidedly.

"Q. Dr. Wells, who was called in this proceeding by the respondent has criticized the work of the Referee Board on several grounds. I think his first criticism was that the Board made its tests upon normal young men and did not include in its tests others who might more readily yield to the operation of any deleterious effect of the food. What have you to say concerning that criticism?

"A. I have had considerable experience in the study of metabolic problems over a period of forty or fifty years, and have visited the laboratories of nearly all the larger European universities where such work is carried on. I think I am justified in saying that experience shows it would be very unwise to select as subjects of such experiments people who are in abnormal condition, excepting where concerned with some specific problem bearing on that particular disease. You are introducing so many variables that your whole problem becomes complicated to a degree which is undesirable and unnecessary.

"Q. Dr. Wells also criticized the work of the Referee Board upon the ground that the feeding experiments were conducted over too short a period to give any substantial ground for making conclusions.

"A. I can merely give my personal opinion that an experiment which extends through a period of one-hundred and thirty days, as my own did in this particular instance, with the feeding of the suspected substance in fairly large doses during that entire period every day, is quite adequate to bring out any harmful results which might follow.

"Q. I think you have already answered this next question, but I should like to ask you once more. Dr. Wells stated in his testimony that one of the effects which he believed would follow from the use of aluminized food would be to rob the body of phosphates. What is your opinion on that subject?

"A. I cannot see any justification for such an opinion. In my own experiments where the dosage was fairly large there was absolutely no change in the excretion of phosphorus in the urine. The level was the same as before. In none of the experiments recorded in Dr. Long's and Dr. Taylor's reports was there anything which would justify such an assumption, except, of course, with very large doses, as in Dr. Taylor's report. Here one might raise the question as to whether such a slight deviation from the normal path might not occur; but I do not find any results which would justify the belief that such a result is ordinarily to be expected.

"Q. Did you find in your work with the Referee Board, or have you since seen, any evidence of the absorption by the body of any aluminum from a diet of aluminized food?

"A. That particular problem was one which interested us very much and the Board decided to have Dr. Taylor give special attention to the question. He reported to the

Board that he had tried the experiment in what he thought was a very thorough manner. Taking 4 men as subjects, he gave them aluminum day by day to the extent of 1 gm. per day, and then he withdrew from their bodies about 200 cc. of blood, making, as he said, a total of more than 800 cc. That would be roughly a little less than 2 lb., the specific gravity of blood being greater than that of water. He subjected that blood to very careful chemical analysis and failed to find any aluminum.

"I think we all understood that owing to the extreme difficulty of making accurate determinations of aluminum (for aluminum is very hard to determine quantitatively when the amount is small), there might have perhaps, been a trace; but he did not find it in that amount of blood; and 200 cc. is certainly a large quantity, drawn from people who had introduced into their systems a very large amount of aluminum.

"Then I have been interested in observations made by other people. One series of experiments or observations which were made in Chicago interested me particularly, namely, the work of Dr. Thorek.

"Dr. Thorek is the surgeon-in-chief of the American Hospital, Chicago, Illinois. In an article* by him on aluminum potassium nitrate in the treatment of suppurative conditions he states that the toxicity or rather the non-toxicity of the aluminum potassium nitrate compound was tested first on guinea pigs, then on rabbits and finally on monkeys by the intravenous, subcutaneous and oral methods, the results in all instances showing that the aluminum potassium nitrate compound was non-toxic irrespective of its method of administration. Two large monkeys, weighing 42 lb. and 45 lb., respectively, were each given single intravenous injections of 80 grains, and later the same amount was introduced intra-abdominally and intramuscularly without the least untoward effect. These animals were at the same time fed 2 oz. each of the salt at each feeding with no loss of weight or decrease in activity or function.

* Thorek, M. Aluminum potassium nitrate in suppurative conditions. *Ann. Surg.*, 1923, lxxvii, 38–47.

"Then comes the point more directly connected with your question: they used this salt as a local application to ulcerated surfaces, employing it as a dressing in large amounts.

"Thorek remarked that in almost every case a marked improvement in the general physical condition of the patient was noted almost at once; and then stated, and this is the point of importance for us, that investigations by Professor Kahlenberg of Wisconsin University failed to disclose any trace of either base in the blood or urine.

"In other words, with this large quantity of this double salt of aluminum potassium nitrate, he apparently found no traces of absorption of the base aluminum or of the base potassium.

"In Schmiedeberg's Pharmacology (and Schmiedeberg is perhaps the foremost German pharmacologist), in the seventh edition, 1913, occurs the following statement:

"'Experiments made by Plagge and Lebbin for the Prussian War Ministry, ten rabbits, ten days to two and a half months receiving daily in their food one-tenth to four-tenths of a gram of tonerde, that is the oxide of aluminum, the earth, in the form of sodium aluminum tartrate, remained completely well, and with the exception of two cases no trace of the base was found in the urine.'

"Again on page 605 he states:

"'Even by subcutaneous application and even by injection into the blood, aluminum passes into the tissue only very slowly.'

"Again on the same page:

"'Absorption of tonerde by intact stomach and intestinal walls, even by giving of aluminum compounds, appears generally not to follow.'

"This is in harmony with the general idea borne out by our experiments: that aluminum is not readily absorbed.

"I should hardly think it wise to treat, in broad generalization, results obtained by the direct immersion of a naked cell, to use your phrase, in a solution of such a substance as alum. One must remember that in alum the action of a soluble aluminum salt like aluminum sulphate or aluminum chloride would obviously be different from what would be obtained by the application of a relatively insoluble compound of aluminum like the oxide or hydroxide or phosphate. But in studying the action of any substance on an isolated cell, or a naked cell, while it is exceedingly interesting and in many cases very important, it is of questionable wisdom or justice to apply the result obtained to what might occur, for example, in the human body.

"The naked cell is one thing and a cell in its natural environment in the tissues or organs of the body is another. In the latter instance there are protective influences at work which cannot rightfully be ignored; what I have already spoken of as the primary and secondary cell constituents must be considered, and also the substances which are being constantly formed by the natural breaking down of the cell in its life changes . . .

"Taking into account the widespread distribution of aluminum and the fact that it is all about us in dust, it would not be at all strange to find traces of aluminum in almost any part of the body. However, such aluminum would be very difficult to detect, and I do not think we know whether there is normally a trace in the blood or not. I should not feel competent to say. I should rather expect on general principles that there is liable to be a trace; perhaps it is from the soil or perhaps taken in as dust, but I do not think we know.

"Q. Assuming that there were found by these investigators minute quantities of aluminum in the blood of the subjects, what have you to say with regard to the ability of the body to dispose of those quantities?

"A. I should not think that quantities such as you have referred to, small fractions of a milligram or even a milligram, if actually present in the blood would have any real significance. Assuming that there is present in the blood a small trace of aluminum, the question is whether or not that aluminum is gradually accumulated in the body, and whether or not one can look sooner or later for a cumulative effect which might be injurious. We do not find any evidence of deposited aluminum; nor of elimination of aluminum through the urine, for example, as we would expect to find if there was any absorption to any degree.

"Q. It would follow from that, would it, that even if the experimenters did find minute traces of aluminum in the blood, there would not be any resulting deleterious effect upon the body itself?

"A. I should not consider that such a finding could itself indicate any deleterious effect. If, however, it were combined with other results, then we might put the two together and such a result would become significant.

"Q. You mean, other results upon that subject where you found the aluminum?

"A. Yes. In forming opinions of this sort, I have to balance such a possibility with the fact that a group of 8 men, living under my observation for four months, and taking relatively large doses of aluminum salts day by day, remained in perfectly good physical condition, with nitrogen balance unimpaired, with no gain in the elimination of nitrogen through the urine, with no change in the utilization of protein food, with no change in the utilization of fat and with a medical examination that showed, according to the words of the medical examiner, perfectly normal physical condition, no change in heart action, no change in peristalsis; all of which implies that there has been no change in the normal condition of the body, no change in functions.

"Q. Is it true that the body is equipped to take care of minute quantities of a foreign substance which may get into it?

"A. If by foreign substances you mean non-utilizable substances, I should say yes, very decidedly. Illustrations are plenty. A large portion of the food that we take into the system is not utilizable, but it is handled thoroughly by the body and with benefit to ourselves. We eat bran with our oatmeal, and the body cannot utilize practically any of it. That is what we call refuse. We eat celery and similar foods, and a large proportion of the material is non-utilized and passes through the intestines with no harm.

"Q. Even if minute amounts of some non-utilizable substances do get into the blood, can the blood dispose of them?

"A. The blood can dispose of them in a variety of ways. They may rapidly be eliminated through the bowel into the intestine, or they may be thrown out through the kidneys: they may be oxidized in the body if they are oxidizable substances.

"Q. If we assume that part of the residue of aluminum salts prepared with an s.a.s. powder is soluble in the gastric juice, is it necessarily absorbed from the gastrointestinal tract?

"A. Not necessarily; not everything that is soluble is absorbed. The absorption is not a purely chemical or physical chemical process. It is a physiological process in which the active agents are living epithelial cells exercising selective action, exercising the power to reject if they see fit; thus the mere fact of a substance being in solution does not necessarily mean that it will be absorbed in the gastrointestinal tract.

"Q. Applying that to our particular problem, was there any evidence in the Referee Board work or in any work concerning which you have since read, that any of the soluble aluminum in the gastrointestinal tract (provided there were such there) was absorbed?

"A. Not to any appreciable extent.

"Q. Assuming that the existence of acidity in soil tends to inhibit plant growth in such soil, and assuming further that the presence of soluble aluminum salt in such acid soil tends still further to inhibit plant growth, would the knowledge of that fact cause you to change your opinion of the effect upon the human body of the ingestion of food prepared with a so-called alum baking powder as stated in the Referee Board report upon this subject?

"A. It would not, because of the great diversity in the conditions. In the one case we are dealing with a group of plant cells, and in the other, with the human body with its complexity of cells and different order of structure.

"There is one statement I should like to make, in order that my position may not be misunderstood. We have been speaking so frequently of alum and alum solutions and of the residues from the baking powder bread with which our experiments have been mainly conducted, that I should like to have on the record a clear statement as to my position with regard to alum or alum solutions; and I have taken a statement from Professor Cushny's book.[6]

"Professor Cushny is professor of Pharmacology at the University of Edinburgh and is, I think I may say, one of the foremost authorities on the subject of pharmacology and therapeutics. The statement I want to read expresses practically my own opinion with regard to the action of alum in alum solutions as distinguished from residues in baking powders.

" 'Alum solutions have a sweetish astringent taste and in small quantities induce no symptoms except a feeling of dryness and astringence to the inner mouth and throat. Larger doses act as gastric irritants and cause nausea and vomiting, and in extreme cases, poisoning.

" 'Even the largest quantities, however, are followed by no symptoms except those of gastrointestinal irritation and inflammation and the long-continued use of alum does not elicit any symptoms of chronic poisoning.

" 'Aluminum salts are absorbed only in small quantities from the stomach and intestines so that no symptoms of general poisoning arise from the internal use of the salt.

" 'Aluminum vessels may be used for cooking or even contain the acid without danger of intoxication, as has been shown by a recent series of investigations.'

"This summation by Dr. Cushny, I believe is a perfectly accurate and well-defined statement of our knowledge with regard to the matter of the reaction of alum. It fits in perfectly with my views as a result of my own experience and experiments."

With reference to experimental studies on the relation of aluminum salts to plant life (Chapter VIII), and on the action of aluminum compounds on unicellular animal life and on the isolated cells and tissues of higher animals (Chapter IX), while Professor Chittenden found them very interesting and important, he was of the opinion that they did not have any particular bearing upon the presence of aluminum salts in baking powder. For example, referring to Dr. Heilbrunn's work (p. 183) he stated: "I feel, however, very strongly that this line of experiments does not have any bearing whatever on the possible action of baking powder salts or residues." Referring to results from the experimental intravenous administration of aluminum salts to animals (Chapter XIII), it was his opinion that such results also were inapplicable.

Referring to a criticism by Dr. Wells of Dr. Taylor's work in connection with the Referee Board report, Prof. Chittenden stated:

"It seemed to me that the criticism was uncalled for and should have been tempered to the facts in connection with the problem as presented to the Board. I will call attention to Paragraph 3 of the questions submitted to the Referee Board:

" 'If aluminum compounds be mixed or packed with the food is the quality or strength of such food thereby reduced, lowered, or injuriously affected; in large quantities; in small quantities?'

"The whole object of some portions of the experiments, as conducted by Dr. Taylor and also by Dr. Long, was to answer that question without reference to the conditions which exist when aluminum baking powders are used in conjunction with the leavening of bread.

"Our understanding with the Secretary of Agriculture was that the primary object was the effect of alum baking powders on the human body when introduced as food in the leavening of bread, biscuits and pastry, but these questions were formulated in a somewhat peculiar way, and we felt it desirable for some members of the Board to carry on experiments with the aluminum compounds not baked in the food, therefore Dr. Taylor tried experiments with aluminum hydroxide emulsified with milk and taken in that form. In certain other cases aluminum hydroxide was mixed with water only. In certain other cases the hydroxide prepared from aluminum chloride was used.

"There was another point which the Referee Board had in mind, and which we all, I think, felt was very important. We were appointed to find out if possible, whether it is true that aluminum compounds have any injurious effect when taken as they are consumed in baking powders, mixed with bread etc.; but we felt that it was our duty to go to the extreme and study the effect of aluminum compounds under conditions perhaps not as favorable. It was for this reason that Dr. Taylor broadened the scope of the work and broadened the field of our inquiry by studying the action of aluminum compounds not combined with the bread and the dough etc., but introduced directly."

VII

LITIGATIONS RELATIVE TO BAKING POWDERS CONTAINING SALTS OF ALUMINUM

NORFOLK BAKING POWDER CASE[1]

The Norfolk Baking Powder Case was a prosecution in England, about 1879, for the sale of a baking powder marketed under that name. The following résumé of the testimony of witnesses, abstracted from the accounts then published, gives a general idea of the reasons then held for the opinions for and against the healthfulness of baking powders containing alum. This was before the era of s.a.s. as the acid ingredient of baking powders.

James West-Knights,[2] public analyst's analysis of the powder: Ground rice, 41.5 per cent; burnt or dried alum, 15.76 per cent; bicarbonate of soda, with traces of potash, silica and moisture, 42.74 per cent. A teaspoonful per pound flour would contain 23 grains of burnt, equal to 44 grains crystallized alum. It forms phosphate of alumina with the bread phosphate, leaving only a trace as hydrate and makes the bread indigestible; hardens the gluten. It is injurious to health.

Matthew Monorieff Patteson Muir, praelector of chemistry: Agreed generally with the last witness. Found that whereas soluble phosphates were generally present, in bread made with this powder there was only a small quantity of soluble phosphates. Would give no opinion as to whether it was injurious to health.

Dr. John Buckley Bradbury: The baking powder would rob the system of soluble phosphates. The only origin of phosphorus found in the nervous system is these phosphates. It would harden the gluten and render it indigestible. The constant taking of such food would be detrimental to health.

Francis Sutton, public analyst: Found $3\frac{3}{4}$ grains Al_2O_3 in a 2 lb. loaf of bread made with the powder; found $\frac{3}{4}$ grain in yeast bread from same flour. His experience supported Prof. Patrick's experiments that it had no effect whatever on the human system.

Dr. Michael Beverly: Was of the opinion that there was nothing in the baking powder, used as directed, injurious for food.

The magistrates found the defendants guilty.

APPEAL. Testimony of witnesses before the Recorder of Cambridge

J. W. Knights: Bread made by this powder would be deprived of seven-tenths of its phosphoric acid and so would be less nutritious. A great many articles of food contained phosphoric acid. Had never heard of indigestion from eating such bread but would judge that indigestion would follow.

M. M. P. Muir: When the powder was mixed with phosphate of soda in place of flour, insoluble phosphate of alumina was formed. Flour free from alum contained a large quantity of soluble phosphoric acid; when half a pound was mixed with half a teaspoonful of the baking powder, it contained very small quantities. Bread made with the powder contained $1\frac{1}{2}$ grains soluble phosphoric acid per pound flour; with yeast, four times that amount. The solubility in 0.2 per cent HCl at 100°F. was determined. When a mixture of gluten and baking powder was digested in 0.2 per cent HCl for fifty hours, 20 per cent less of the gluten was dissolved than without the powder. In a mixture with the baking powder, dextrine was precipitated with the aluminum phosphate.

Dr. J. B. Bradbury: Regarded the powder unfavorably because it rendered the phosphate insoluble. People suffered from indigestion who did not eat bread. If the phosphate of alumina were not insoluble in the stomach it would modify his opinion.

Mr. Blofeld: This baking powder had been in use thirty-nine (?) years and many millions must have eaten bread made with it.

Dr. Paget, professor of physics, University of Cambridge: On the basis of previous testimony thought that from eating an ordinary amount of bread made with the powder in the course of time digestion would be impaired, and this would be more marked with children and invalids. Probably one took more phosphoric acid per day than was good for one. His opinion was based upon the fact it rendered gluten and dextrine less soluble.

Francis Sutton: Believed hydrate and not phosphate of alumina was found in bread. If phosphate, it would be soluble in the gastric juice. It was a weak chemical combination; many fluids would decompose it. Yeast bread gave him 3.04 grains phosphoric acid, per 4 lb. loaf, dissolved in water; Norfolk baking powder, 2.32 grains. From his experiments he did not agree that alumina rendered the gluten less soluble. He mixed the phosphate with gluten in very large excess. Solubility in hydrochloric acid was not a fair test, since pepsin is present in gastric juice and there was the admixture with the salivary fluids.

Two pigs fed on bread from 91 lb. of flour with the Norfolk powder for eight days were killed at the end of twenty-eight days and the contents of the stomach showed the mixture of the powder. The feces from the upper and lower bowel were examined. The upper bowels showed an excess of aluminum over phosphoric acid of 2.07 per cent; the lower bowels 3.15 per cent. The inference was that the gastric juice took all the phosphoric acid it required, leaving the hydrate of alumina to be rejected with the feces. The internal appearance of the pigs was perfectly healthy.

He did not think 40 grains of alum per 4 lb. loaf of bread injurious, but it should not be allowed, as it opens up the way to fraud.

He certainly could say as a skilled witness, that there was nothing injurious in this powder.

Dr. J. L. W. Thudicum: The diminution of the phosphoric acid in the human body by the use of this powder would be inappreciable and would be of no consequence to the body.

Dr. Charles M. Tidy, professor of chemistry and forensic medicine, medical officer of health, etc.: He thought it was very improbable that phosphate of alumina was formed in bread; even if formed, he did not think it made the slightest difference. It was soluble in the gastric juice and filtered through into the blood. As to whether he could see in the use of this powder anything injurious to health, no, most certainly not.

The Recorder: "After the evidence we have just heard, I do not think this baking powder is an article of food or that bread made with it becomes an article of food injurious to health, and, as a matter of fact, I find in favor of the appellants."

It will be noted that the reasons given in support of the opinions that the powder was injurious to health were:

1. That the aluminum formed phosphates of aluminum which were insoluble in the gastric juice, hence not absorbed.

2. That, because of (1), the body was thereby deprived of needed phosphates.

3. That the gluten of the flour was toughened and both gluten and dextrine were rendered less soluble and thereby indigestible.

CASE, ABOUT 1892[3]

About 1892, Otto Hehner, President of the Society of Public Analysts, brought a case into court, in which evidence favorable to a powder containing aluminum was offered by Wynter Blyth, author of work on foods, and F. Sutton, public analysts, both of whom condemned alum in flour, and by Dr. Thudicum, who considered alum itself to be possessed of valuable digestive properties. The magistrates dismissed the case on the same grounds enunciated by the Recorder of Cambridge in the Norfolk Baking Powder Case in 1879.

EXCELSIOR BAKING POWDER CASE[4]

In February, 1893, an action was brought against James James, a grocer, and he was convicted for selling "Excelsior" baking powder, containing "an injurious ingredient." The case was heard on appeal, April 12 to 17, 1893, before Glamorganshire Quarter Sessions.

Dr. W. Morgan, public analyst, certified that the powder contained 39 per cent alum; 360 grains per 4 lb. loaf would yield aluminum hydrate to the amount of 6 grains per lb., soluble in the gastric juice as aluminum chloride which was noxious to the stomach. Dialysis of stomach contents obtained by inducing vomiting showed the presence of aluminum chloride. The aluminum hydroxide was not mixed with the bread, but was wrapped in a muslin bag and introduced into the dough.

Mr. Otto Hehner regarded the powder as injurious by reason of the solubility of the hydrate of alumina in the stomach.

Claude Thompson, professor of chemistry, gave substantially the same evidence.

Wyndham Dunstan, professor of chemistry, had found the hydrate of alumina soluble in gastric juice and also that it interfered with the digestion of starch (diastase) and with peptic and pancreatic digestion. Moreover, 0.3 per cent carbonate of soda dissolved hydrate and phosphate of alumina, so that absorption would occur. This was confirmed by finding alumina in the urine of a man who had daily taken 15 to 30 grains of the hydrate. On one occasion it had induced vomiting.

Dr. Lauder Brunton testified that alum caused disturbance of the digestive functions. He regarded the presence of hydrate of alumina in the stomach as injurious.

Rhymer Marshall, D. SC. testified to the formation from the hydrate of aluminum chloride in the stomach.

Dr. Thomas Druslyn Griffiths stated that the effects were both local and constitutional, that they were highly irritating to the internal organs.

Dr. Ebenezer Davis, medical officer, submitted that 6 grains of hydrate of alumina retarded the progress of digestion.

Dr. Thomas Henry Morris stated the alum in "Excelsior" baking powder would be highly injurious to health.

Francis Sutton, public analyst, had used "Excelsior" baking powder in his family for upwards of thirty years (?) without ill effect. He did not believe hydrate of alumina, as it occurred in bread, soluble in the gastric juice. He had administered a pound of bread made with the baking powder in question in the amount indicated, also $1\frac{1}{2}$ pt. water and had obtained the stomach contents after $1\frac{1}{2}$ hr. There was no aluminum diffused through the walls of a sheep's bladder (Dr. Morgan used parchment paper).

Dr. Arthur Pearson Luff, as a specialist on poisons, declared hydrate of alumina an inert substance and entirely harmless. He did not believe it soluble in the gastric juice, basing his opinion on the insolubility in .02 per cent (?) HCl. He did not consider Dr. Morgan's experiments fair because of the amount given and the hydrate was not baked mixed with the flour of the bread.

Dr. Benjamin Ward Richardson had used "Excelsior" baking powder in the bread he had eaten for the past three weeks and in his opinion it was not at all injurious to health. He had never known of any ailment from the use of these powders taken with impunity by hundreds and thousands of people.

Ten lay witnesses and several physicians testified they had used the baking powder in their homes without apparent injury to health.

Mr. Wynter Blyth, medical officer, analyst, and author of works on hygiene, testified that he had used a similar powder and did not regard it as injurious to health; that the hydrate of alumina in Dr. Morgan's experiments was freshly precipitated and excessive in amount.

The Court confirmed the conviction of the lower Court. In the discussion following the decision, the appellant asked for the sanction of the court to points practically agreed upon between respondent and appellant, and that "a case should be granted upon the following points":

"1. That there is no evidence that any article mixed with any ingredient or material injurious to health, was sold, and that baking powder is not an article of food within the meaning of the Food and Drug Act, 1875, Section 3.

"2. If baking powder is an article of food, then there was no evidence to show that any ingredient injurious to health was mixed.

"3. That there was no evidence that the baking powder actually sold was injurious within the meaning of Section 3."

The Court could not agree to Point 3: "It is a matter of fact." When reminded by counsel that no evidence of any specific injury had been adduced, the Court replied, "But if we find it was injurious, there must have been evidence that it must injure to some extent." Counsel: "But there never has been." Court: "But we find it."

In this case, the claims upon which were based the opinions of witnesses that the baking powder was injurious to health were:

1. That the alum baking powder would be noxious to the stomach, since, when aluminum hydroxide was mixed with dough and the bread therefrom subsequently eaten and the stomach contents dialyzed, aluminum was found in the dialyzate.

2. That aluminum would interfere with diastatic, peptic and pancreatic digestion.

3. That after the ingestion of 15 to 30 grains of aluminum hydroxide per day, aluminum had been found in the urine of a man.

4. Aluminum was highly irritating to the internal organs.

The opinions of witnesses favorable to the baking powder were:

1. That the aluminum residue from baking powder was not soluble in the gastric juice.

2. That the use of the baking powder by hundreds and thousands of people was never known to have produced injury to health.

STATE OF MISSOURI VS. WHITNEY LAYTON[5]

In the year 1899, the General Assembly of the State of Missouri enacted the following law:

"Section 1. That it shall be unlawful for any person or corporation doing business in this State to manufacture, sell or offer to sell, any article, compound or preparations, for the purpose of being used, or which is intended to be used, in the preparation of food, in which article, compound or preparation, there is any arsenic, calomel, bismuth, ammonia or alum.

"Section 2. Any person or corporation violating the provisions of this act shall be deemed guilty of a misdemeanor and shall, upon conviction, be fined not less than one hundred dollars, which shall be paid into and become a part of the road fund of the county in which such fine is collected.

"Approved May 11, 1899. Takes effect August 22, 1899."

To test the constitutionality of this law, as applied to baking powders containing S.A.S., Whitney Layton was charged with manufacturing and selling in the city of St. Louis, Mo., on August 28, 1899, a can of baking powder containing S.A.S. as an ingredient. In the trial of the case, this fact was admitted, and the question was not raised whether S.A.S. was "alum."

George C. Rew, chemist, testified that no free alum was left in bread prepared with a properly made S.A.S. baking powder and

that an excess of s.a.s. over and above the amount required for
the liberation of gas from the sodium bicarbonate would not be
used, especially since it is the most expensive constituent in these
powders. Also that his company did a very large business and that
no complaints had ever been received that food prepared with
their powders was in any way injurious.

Five manufacturers or wholesale dealers of s.a.s. baking powders
testified that they had never had complaints nor did they know
of any unwholesome effects from food prepared with their products.

Eleven operatives employed in manufacturing s.a.s. or in
factories making s.a.s. baking powders, all of whom had inhaled
daily over long periods of time air laden with s.a.s. dust, testified
that they had never had any injurious effects from such dust nor
did they know of any such injurious effects upon their fellow
workers, who totaled a very large number.

Ten managers of hotels and keepers of boarding houses testified
that they had used s.a.s. powders in their institutions over long
periods of time, that such powders had proved satisfactory, that
they had never had complaints of unwholesomeness of food prepared with the powders and that they had never known of any
injurious effects.

Dr. E. E. Smith testified in detail the conduct of experiments
by him (see pp. 61–69) and concluded that a bread properly made
with a so-called alum baking powder was perfectly wholesome.

Dr. Austin Flint, the physiologist, testified that he considered
that the questions involved in the matter of the nutritive value
and wholesomeness of food prepared with a so-called alum baking
powder were essentially physiological rather than chemical and
that to his knowledge, prior to the investigations by Dr. Smith,
there were no connected investigations of the subject. He had
followed the investigations by Dr. Smith and had heard Dr.
Smith's testimony in this case and in his opinion the experiments
were carried out properly, accurately and in every way were
entitled to confidence. Assuming Dr. Smith's testimony as to the
facts to be true and looking at the question from what he believed
to be a purely disinterested scientific point of view, he regarded
the bread designated as the Layton bread unquestionably to be
a wholesome bread used as an article of food. He regarded such
bread as not differing from any other bread as regards its whole-

someness for various classes of consumers, those of strong digestion or weak digestion. In his experience as a physician, he had never been led to attribute any functional disorder or diseased condition to the use of "alum" baking powder in the preparation of food.

Dr. Albert Merrell, a practicing physician in St. Louis, Mo., for twenty-five years, twelve years a member of the Missouri State Board of Health and six years a member of the St. Louis, Mo., Board of Health, testified that he had never known of any case where the digestion or health of a person had been injuriously affected by food made with an "alum" baking powder and that he did not believe such a food would have any injurious effect.

Similar testimony was given by Dr. Degrange Atwood, a practitioner of forty-nine years' experience and by Dr. George Homan, for several years a member of the Missouri State Board of Health, for two years Health Commissioner of St. Louis, Mo., and professor of hygiene in a St. Louis Medical College.

To disprove the conclusions of the foregoing witnesses that food is not rendered unwholesome because it is prepared with a so-called alum baking powder, the manufacturers of a tartrate powder put in evidence for the state the testimony of witnesses including several eminent chemists and physicians.

Prof. J. W. Mallet gave in detail the results of investigations already published (see pp. 51–52) and reiterated his conclusions announced in that article.

Further, he referred to the injurious action of aluminum salts when administered subcutaneously (see Siem pp. 47–49 and Döllken pp. 56–58). He likewise suggested that baking powders might be made with the wrong proportion of ingredients, that the ingredients of the powder might not be thoroughly mixed, that the baking powder might not be thoroughly mixed with the flour and that because of deterioration of the powders larger amounts than the indicated quantities might be needed for the aeration. He denied that aluminum was a natural ingredient of plants, explaining its seeming presence by contamination of the plant material analyzed by adherent soil. Although he testified that aluminum has very little tendency to accumulate in the system, he was of the opinion that the continued ingestion of quantities too small to produce demonstrable effects would eventually work an injury. He knew of no instance in which injury had been produced by the aluminum of baking powder food.

Prof. A. B. Prescott had not conducted any experimental work on food prepared with "alum" baking powder. Besides agreeing with some of the opinions given by Professor Mallet, he mentioned the possibility that a basic sulphate of aluminum might be formed instead of the hydroxide during the decomposition of the baking powder. Also that the aluminum was soluble to some extent in the acids of gastric juice and that the presence of sugar and other soluble carbohydrates in the digestive juice might render the aluminum hydroxide somewhat soluble, even when the acids of the gastric juice were neutralized in the duodenum.

Prof. Victor C. Vaughan had not made any experiments to determine the effects of food made with "alum" baking powder upon the animal or human system. His conclusion that such food was injurious was chiefly if not wholly based upon the work of Siem (pp. 47, 48, 49), who studied the effect of aluminum salts upon animals when injected subcutaneously. Hence, he testified: "They produce what is called bulbar paralysis. The anterior horns of the spinal cord finally begin to degenerate, and change manifests itself, both motor and sensory. The patient loses both sensation and motion. This gradually extends, as the degeneration goes on, until finally the whole of the spinal cord is involved, and lastly the brain. At about this time death ensues. Salts of aluminum also produce what is known as the "metallic" kidney and a fatty degeneration of the muscles of the heart, which also in time lead to death. These results follow in a few times, or when small doses are given repeatedly for a long time."

The following interrogation by the court indicates the basis of the witness's opinions:

The Court: "Doctor, I would like to ask you a few questions regarding this matter: Assuming that the patient was a person of normal constitution and should take bread baked by the use of alum baking powder which contained, as a residue from the baking powder, aluminum salts in the quantity you have stated in your testimony, say for a period of months and years, what in your opinion would be the effect of those aluminum salts so taken, on the general health of the person and the digestion of the subject?"

A. "I don't think you could lay down any general law applicable to everybody as to what extent the salts of aluminum are absorbable through the intestines; therefore, it would be hard to state the effect of the general proposition."

Q. "But that is just the whole question. I have presupposed a person of normal health, what you would call an average person."

A. "I do not know because I do not know how much of the aluminum salts in the bread would be absorbed. That would depend upon the amount of lactic acid that would be present and the amount of soluble absorbable compounds with the protenoids. I may state that it will practically be impossible to state the exact effect under the circumstances detailed in your question."

Q. "You say it would be impossible to tell just what the effects would be?"

A. "It would be impossible to tell under those circumstances."

Q. "Do you mean to say that therefore the salts would have no ill effects, or what is your testimony on that point?"

A. "No, I didn't state that. My testimony is that it would be impossible to tell under those circumstances what the effects would be. There might be some disturbance of the digestive functions."

Q. "What, in your opinion, would be the effect under those circumstances?"

A. "Well, I must confess that I cannot answer the question because in some cases it might not have the slightest ill effects and in other cases there might be some disturbance of the digestive functions. It would be impossible to lay down any general proposition governing those cases."

Q. "In your opinion would any human being be affected injuriously or otherwise by it?"

A. "I think in all probability he would."

Q. "In what manner would such person be affected?"

A. "From what we know from experiments upon animals a person would probably slowly develop some disease of the spinal cord. That would most likely be the result."

Q. "Has there ever come under your observation any case of that kind during your twenty-five years of practice which you attributed to that cause?"

A. "I have seen cases for which I could not assign any cause.

Q. "Do you mean by that then that you would assign the use of baking powder containing alum as the cause?"

A. "No, I would not like to say that."

Q. "Did you in all of your twenty-five years of experience ever have a case of that kind which you attributed directly to the use of alum in baking powder?"

A. "No, sir; I cannot say that I ever had such a case."

Q. "In the course of all your reading and getting information from books of science on this question have you ever seen recorded a case such as you have described which would be attributed to the use of alum baking powder in the preparation of food?"

A. "No, sir."

Q. "Then your opinion as to the deleterious effects of alum baking powder is based upon theoretical propositions?"

A. "Entirely theoretical and experiments upon animals from which we judge of the effects on the human system."

Q. "Then your statements are not based in any wise upon cases that have come under your individual observation or that you have seen recorded in books?"

A. "No, sir; from experiments on animals."

Criticisms by this witness of the experiments of E. E. Smith are considered in the published record of this work, quoted on pp. 63, 68.

Dr. Paul Schweitzer, professor of chemistry, State College of Missouri, testified that he had obtained alum from a dilute hydrochloric acid extract of bread made with an s.a.s. baking powder and that alum was present in the food because it gave a coloration with tincture of logwood. He opposed the use of any food containing alumina or sodium sulphate; would rather drink mud than water purified with alum.

Dr. William B. Potter, qualifying as an expert chemist, mixed a sample of s.a.s. baking powder with water, collected the insoluble residue and heated it in a bag in the center of bread dough baked in the usual way. He testified that subsequently, 62.5 per cent of the alumina of this residue dissolved when exposed to 0.25 per cent hydrochloric acid for two and one-half hours at a temperature of

100°F. s.a.s. baking powder bread similarly yielded 0.05 per cent aluminum hydroxide to 0.25 per cent hydrochloric acid. For these reasons he regarded food made with s.a.s. powders unwholesome. He did not know of any illness or injury caused by eating such food.

Dr. William S. Moore, a medical practitioner residing in St. Louis, regarded food prepared with an s.a.s. powder injurious for the reasons put forth by the previous experts. On theoretical grounds he believed it would diminish the gastric juice and produce gastrointestinal catarrh, indigestion and constipation. He had made no experiments nor had he seen or read of any case where injury had resulted from eating such food.

Dr. Alonzo R. Kieffer, professor of anatomy in the Barnes Medical College and a medical practitioner in St. Louis, and Dr. James A. Chase, professor of pathology in the Barnes Medical College and a medical practitioner in St. Louis, each gave testimony similar to that of Dr. Moore and neither had known or heard of injury or disease from eating s.a.s. baking powder food

Dr. Ludwig Bremer, a medical practitioner in St. Louis, similarly testified, basing his unfavorable opinion more particularly upon the action of alum as such in food. He had administered 2 to 10 grain doses of alum for continuous periods and observed indigestion and constipation.

William M. Chauvenet, a chemist, had digested bread made with the same powder used by Dr. Potter and obtained 25.9 per cent of the alumina in solution in artificial gastric juice.

The opinion rendered by the Court stated:

"The prosecution contends this substance to be deleterious, and the eminent experts testified on that subject as to the theoretical and general knowledge of these matters that serious results would follow the consumption of food containing alum.

"Professor Mallet, of the University of Virginia, presented reports of experiments conducted by him upon himself as a subject, and his report was that, taken in single doses, it required not less than 20 grains of hydroxide of aluminum to produce any appreciable effect upon the subject. All other experts who testified for the prosecution testified purely upon a theoretical basis, and, although it appears to the Court that hydroxide of aluminum is a substance easily accessible, and which could have been made the subject of practical experiments whereby the various eminent scientists could have made tests, yet they were without a single practical test, with the exception of that of Professor Mallet, and were without a basis of actual determination upon which to found their theory.

"Upon cross examination the experts testifying for the prosecution admitted that in all their experience and in all their reading and information that they possessed on the subject they had never themselves come in contact with, nor could they obtain any information or any knowledge of any recorded instances in which functional disorders or disease or impairment of the digestion and general health had resulted to any human being from the use of alum baking powder as an ingredient in the preparation of food.

"In the mind of the Court this fact, considering the enormous proportions to which the alum baking powder industry has grown in this country, and the length of time in which such baking powders have been in use, stands as a stone wall against the deductions of the most eminent scientists who presented their theories on the part of the prosecution. I am unable to find in the evidence in this case any just ground for a ruling that alum baking powders, of themselves, when used in the preparation of food are in any wise less wholesome than any other variety of baking powders. This finding, however, does not or in its nature cannot dispose of this case. . . . It appears from the agreed statement of facts that the act denounced by the Legislature having been committed by the defendant he must necessarily be found guilty in this court, and a fine will therefore be imposed upon the defendant of one hundred dollars."

On appeal, the verdict of the Missouri Supreme Court was: "The mere wisdom or unwisdom of the act, is not for us to decide. The judgment must be and is affirmed."

The Supreme Court of the United States dismissed the writ of error, because of lack of jurisdiction, since the constitutional question involved was not raised in and submitted to the trial court.

In this case, the reasons given in support of the opinions that s.a.s. baking powders were not injurious to health were:

1. That the acid (aluminum) ingredient of the powder was entirely decomposed by the sodium bicarbonate, leaving merely a small amount of inert residue.

2. That despite the preparation of food with many thousand pounds of powders of this class and its consumption by hundreds of thousands of individuals:

a. Manufacturers and wholesale dealers of such powders had never received complaints of injurious effects to consumers.

b. Managers of hotels and boarding-house keepers had never known of the production of any injurious effects to consumers.

c. Physicians of wide experience in the practice of medicine had never known of the production of any injurious effects to consumers.

3. That experimental studies of the ingestion of food prepared with S.A.S. powders had shown:

a. That there was no interference with gastric secretion.

b. That a diet containing a relatively large amount of food prepared with S.A.S. powders was taken up by the body in the same amount as a control diet in which added aluminum was not present, and further, during the consumption of this relatively large amount of baking powder food there were no digestive disturbances, indicated either by clinical symptoms or by changes in the composition of the urine and stools.

4. That aluminum in varying amounts was a natural constituent of food, hence, the use of S.A.S. as the acid ingredient of baking powders did not introduce a foreign element into the food.

The reasons given for the opinions that food prepared with an S.A.S. baking powder were injurious to health were:

1. The ingestion of aluminum hydroxide, as such, and without accompanying food, produced oppressive gastric sensations, when the amount was much greater than would be ingested at one time with baking powder food.

2. When either aluminum hydroxide or aluminum phosphate was added to pepsin hydrochloric acid solution, 50 per cent, more or less, was dissolved and rendered about 30 per cent of the pepsin insoluble.

3. When aluminum salts were injected subcutaneously into animals, degenerative changes were produced in the nervous system.

4. Theoretically, aluminum salts would diminish gastric secretion and produce gastric catarrh, indigestion and constipation.

STATE OF MISSOURI VS. THE GREAT WESTERN COFFEE AND TEA COMPANY[5]

In 1901, under the act of May 11, 1899, upon which the Layton case was based, an action was brought by the State of Missouri against the Great Western Coffee and Tea Company for selling their own brand of baking powder which analysis showed to contain aluminum. Evidence bearing upon the question as to whether aluminum was unwholesome or injurious was not presented in this case. The defendant company was found guilty in the lower court, but in the Supreme Court of the State of Missouri, on the constitutional ground that the Act was directed against manufacturers and not sellers, the cause was reversed and the defendant discharged.

Subsequent to the Layton and Great Western Coffee and Tea Company cases, the Missouri State law of 1899 as regards alum was repealed and there was no further action against the sale of S.A.S. baking powders.

THE PEOPLE OF THE STATE OF CALIFORNIA VS. (1) J. A. SAINT, (2) A. ZIMBLEMAN[5]

In 1903, under the laws of the State of California, the Board of Health of the City of Los Angeles directed the arrest of two dealers in groceries for the sale of "K.C." baking powder containing S.A.S. as an acid ingredient under the charge that such powders were adulterated. It was contended on behalf of the defendants that baking powder containing alum was not within the restrictions of state laws prohibiting the sale of adulterated articles of food or articles useful in the preparation of food. A separate opinion of the lower court was given in the case of each defendant. In the case of one defendant, the opinion stated:

"Baking powder is defined by the Century Dictionary to be; 'Any powder used as a substitute for yeast in raising bread, cakes, etc. Baking powders are composed of bicarbonate of sodium or potassium mixed with a dry powder capable of setting carbonic acid free when the mixture is moistened.'

"Food is defined to be: 'What is fed upon to support life by being received within and assimilated by the organism of an animal or plant; nutriment; aliment; provisions; victuals.' (Bouvier's Law Dict. Rawle's Ed.). 'What is eaten for nourishment; whatever supplies nourishment to organic bodies; nutriment; aliment; as the food of animals consists mainly of organic substances.' (Cent. Dict.)

"Is baking powder a substance which is fed upon to support life by being received within and assimilated by the organism of man? Is it eaten for nourishment? Does it supply nourishment to the animal organism? Is it nutriment or aliment?

"I think all these questions must be answered in the negative. Baking powder is designed to take the place of yeast, in raising bread, etc. in cooking. This result is attained by the combination of certain substances which, in process of cooking, by chemical action, combine and are thrown off in the form of gas, which permeates the dough and renders it light and digestible. Probably very little of either of these substances remains in the bread or other food when it is cooked; and that which remains, if any, is not in its original state, but is a small deposit resulting from the chemical action. This deposit and what is known as the filler of the baking powder are all that remain in the food. The baking powder is not added to the dough to make the finished product more nutritious, but for the sake of its chemical action, whereby the baking powder itself is, mainly, dissipated, and passes through the dough in the form of gas.

"Hence, I am of the opinion that baking powder does not fall within the definition of an article of food, and this conclusion renders it unnecessary to consider the second point urged by defendant in this case.

"The demurrer will be sustained."

In the case against A. Zimbelman the Court delivered an oral opinion. The question as to whether baking powder containing alum is an adulterated article useful in the preparation of food was considered and decided, the Court answering the question in the negative, and declaring that baking powder containing alum was not adulterated.

COMMONWEALTH OF PENNSYLVANIA VS. MEYER GROSS[6]

In 1910, the Court of Quarter Sessions, Dauphin County, Pennsylvania, rendered a decision relative to the meaning of the word "alum" (see p. 31). In an enactment of the State Legislature, the sale was prohibited of food containing "boric acid or borates, salicylic acid or salicylates, formaldehyde, hydrofluoric acid or fluorides, fluoborates, fluosilicates, or other fluorine compounds, dulcin, glucin, saccharin, alum, compounds of copper, betanapthol, abrastol, asaprol, oxides of nitrogen, nitrous acid or nitrites, pyroligneous acid, or other added ingredients deleterious to health," etc. Under this enactment one Meyer Gross was indicted for selling a baking powder containing s.a.s. as an acid ingredient. During the trial of the case, evidence was offered tending to show that s.a.s. baking powders were not injurious to health. Upon this point, the Court ruled that the freedom of the baking powder from a substance injurious to health was not in controversy in that trial. Hence evidence as to the effect on health was not admitted. The Court did not agree with the contention of the defense that baking powder was not a food within the contemplation of the Act; nor that the use of the word "added" in the Act, in the clause, forbidding alum, simply made an offense where the named injurious or deleterious substance was added to an article that was already complete in itself before such addition.

The principal controversy in the case arose over the meaning of the word "alum" in the Act. During the conduct of the trial, the Court construed the meaning to be that given by the Commonwealth's witnesses, that is, to include 3 double salts of aluminum, namely, potassium aluminum sulphate, sodium aluminum sulphate and ammonium aluminum sulphate with or without water of crystallization, and the single salt, namely, aluminum sulphate with or without water of crystallization. Under the ruling, the defendant was found guilty.

In an opinion upon a motion for a new trial and for arrest of judgment in the case, the Court reversed its ruling upon the meaning of the word "alum" in the act in favor of the contention of the defense, namely, that the word "alum," in the singular, meant that which was known generally to the non-scientific world as alum, which was potassium aluminum sulphate with the water of crystallization, that is, that it was the commonly known crystalline alum which was sold everywhere under that name. This conclusion was based upon the following reasons:

"1. That all of the expert witnesses on both sides concede that alum as commonly known in the United States, to the non-scientific at least, is potassium aluminum sulphate in its crystalline form.

"2. That practically all of the definitions given for the word "alum," singular, in scientific works as well as in dictionaries, speak of potassium aluminum sulphate with the water of crystallization as the common alum of commerce and trade, and when it is intended to refer to other double or single sulphates of aluminum use the qualifying word necessary to express the class or kind . . . It is also clear from the definition given in some of the authorities, that a large number of substances are designated by the general name of alums. There are also double sulphates in which aluminum may be replaced by iron, chromium or manganese, and also in which the potassium may be replaced by some other alkali or ammonium. The general name of alums, therefore, it seems to us, would have a much wider meaning and include a much greater number of substances than was contended for by the expert chemical witnesses of the Commonwealth. . . .

"3. The context of the word "alum," in the fifth paragraph of Section 3, is a strong argument against the contention of the Commonwealth . . . If in all other substances the words used indicate a definite and distinct substance, it is persuasive argument to say that the word "alum" also must have designated a definite and distinct substance. This is especially true in view of the fact that the author of this Act could have so easily removed all doubt as to what was meant. In this case, it appears from the practically uncontradicted evidence, and the jury have so found, that the substance in the baking powder in question is a mixture or compound of sodium sulphate and aluminum sulphate anhydrous, that is, without the water of crystallization. This substance is known to the chemical trade, according to the evidence in the cause, as S.A.S., meaning sodium aluminum sulphate. Had the Legislature intended the construction contended for by the Commonwealth, it certainly would have been easy to have used words clearly covering the point which it now attempts to include. Had the word been used in the plural even, much less doubt as to what was meant could have arisen. The use of the word aluminum sulphates would clearly have covered the substance here used. We conclude, therefore, that the context of the Act sustains the contention of the defendant.

"4. One of the rules of construction which the Courts should have in mind in construing the words of an Act of Assembly is, the evil intended to be remedied. In this case it seems to be conceded on all sides that common alum has been in common use in certain food products. It is fair to assume, therefore, that in using the word "alum" the Legislature had in mind that which was ordinarily known and used as alum. It is admitted by the defense that common or potash alum was formerly used in baking powders, but they claim that it has not been so used for many years. Common alum, it appears from some of the testimony, was recently used to enhance the appearance or value of pickles, and such use may have been aimed at by the Legislature. The testimony in this case is clear that the product used in the baking powder in controversy is a mixture or compound containing sodium sulphate and aluminum sulphate, and is known to the trade as S.A.S. Had this legislation been aimed at this trade, the Act of Assembly ought to have been so worded that the baking powder trade would have notice of the legislative intent. The people affected are entitled to have a hearing before the Legislature as to the injurious or deleterious character of the product intended to be prohibited. The proper tribunal to determine the question of the character and its effect on health is the Legislature. It is very clear from the difference of opinion expressed in the trial of this case that the word "alum" used in the Act could not have been notice to the baking powder people that their business was being affected by adverse legislation. The common, universal and non-scientific use and acceptance of the word "alum" would not include the substance known to the baking powder trade as s.a.s. The definition of the expert chemists of the Commonwealth is purely technical and moreover very much disputed. We are convinced, therefore, that the disputed definition of the word "alum" could not have been known to the Legislature as inclusive of s.a.s. or words would have been used so as to cover the intended substance admittedly used in baking powder. We believe, there-fore, that the Legislature did not have in mind s.a.s. baking powder as an evil to be remedied.

"5. The statute in controversy is a penal statute, which as a rule are to be strictly construed. . . . The true construction of all Acts of Assembly is to get at the meaning of the language used by the Legislature. The words of this Act must be construed with reference to the legislative intent. What then was intended by the use of the word 'alum'? Should it be construed as widely as is contended by the Commonwealth? We think not. No word in a penal Act of Assembly should be the subject of such contradictory testimony. If 'alum' is a distinctive substance, commonly known as such to common and not scientific people, it is fair to assume that that only was meant. The construction widening the word 'alum' to include other and different substances, which may be called 'the alums,' is entirely too indefinite and doubtful in a penal statute. Such construction invokes liberality not to be com-mended in any statute. Legislation looking to the purity of food products should be upheld by the Courts, but where language is used so extremely doubtful in meaning and uncertain in use, as was developed in this case, we are of the opinion that the Courts should, as in all other cases of like nature, give the benefit of that doubt to the defendant. At best not only the testimony but all of the written authority defining alum, consulted by the Court, leads us to doubt whether alum, singular, can be construed to include the substances known as s.a.s. admittedly a constituent part of the baking powder in controversy.

"For the above reasons, therefore, we believe we are in error in our construction of the word 'alum' at the trial, from which it follows that the defendant was entitled to a verdict of not guilty."

On the question of the unconstitutionality of the Act because of the exemption of the retailer or seller holding a guaranty of the manufacturer, wholesale dealer, jobber or distributor, the Court recognized that there was very strong proof for the argument that this immunity clause granted special privilege to the retail dealer and for that reason was offensive against the constitutional inhibition and of very doubtful constitutionality, but in view of its construction of the word "alum," found it unnecessary to rule upon that point.

The order of the Court was:

"Now, June 1, 1910, upon due consideration of the motion in arrest of judgment in the above stated case, the said rule is made absolute, the verdict of guilty is set aside, judgment thereon arrested and the defendant discharged."

FEDERAL TRADE COMMISSION, COMPLAINANT VS. ROYAL BAKING POWDER CO., RESPONDENT, DOCKET 540[7]

On April 18, 1923, the Federal Trade Commission issued a formal complaint against the Royal Baking Powder Company, charging it with the employment of unfair methods of competition in that it had disparaged and defamed goods of its competitors and had falsely charged competitors' baking powders as being poisonous and had published such statements as the following: that they were made from ground-up aluminum cooking utensils; that such competitive baking powders did not come within the pure food laws; that competitors' powders puckered up the stomach in the same manner that alum puckered up the mouth; that competitors' powders were made of the same substance which is used for styptic purposes after shaving. The complaint also alleged that the Royal Baking Powder Company, in addition to these alleged false statements, anonymously disparaged and attacked the wholesomeness of competitors' baking powders by causing the publication of derogatory opinions and statements with regard to the wholesomeness of its competitors' powders, but carefully concealed its connection with such publications.

The Royal Baking Powder Company answered this complaint and in substance denied uttering some of the alleged false statements and justified the publication of other statements by asserting its competitors' baking powder did contain alum and that such baking powders were in fact deleterious to health.

Hearings were held before an Examiner and, after the government had established as a fact that the respondent had uttered and published, for many years, disparaging, derogatory and defamatory statements with respect to baking powders containing sodium aluminum sulphate, the respondent offered in defense a very considerable amount of scientific evidence seeking to prove that its attacks upon competitors' goods were true. Considerable experimental data presented elsewhere in this volume was offered by the respondent and the government rebutted by calling numerous scientific witnesses. The respondent did not attempt to prove that any human being had been deleteriously affected by eating foods prepared with s.a.s. baking powders and, except in the case of some rat-feeding experiments offered by respondent's witness, Gies, the feeding experiments conducted by the witnesses for the respondent developed no evidence of deleterious effects upon animals or man.

This case was presented to the Federal Trade Commission on brief and oral argument in March, 1926, and the complaint was dismissed by the Commission upon its understanding that the attacks of the Royal Baking Powder Company upon the baking powders of its competitors had been discontinued, and that the only thing left in the case for the decision of the Federal Trade Commission was whether or not it was legally unfair for the respondent to assert in its advertising that Royal Baking Powder did not contain alum. The Commission therefore dismissed the complaint, as it found that such a statement with respect to the Royal Baking Powder Company's own baking powder was not legally unfair competition. Counsel for the Federal Trade Commission immediately filed a motion asking that the case be re-opened, alleging that it was not true that the Royal Baking Powder Company had discontinued its defamation of competitors' baking powders and after hearings before the Commission the order of dismissal was vacated and the case re-opened for the taking of additional testimony. At this writing this proceeding against the Royal Baking Powder Company is still pending and it is expected that further testimony will be taken.

VIII

THE RELATIONS OF ALUMINUM SALTS TO PLANT LIFE

VALUE OF SUCH OBSERVATIONS

The relations of aluminum compounds to plant life may have only a remote bearing upon the effect of such compounds on higher animals and on man, when ingested as components of food. Nevertheless, viewed broadly, such relations are of interest because they bear upon the very fundamental question as to whether such compounds are ever biogenic, that is, whether they serve a useful purpose in living processes, or whether they must be regarded as essentially foreign and even essentially inimical when brought into intimate contact with the living cell. If they are without value to life processes and are capable of toxic effects, they must be regarded as essentially injurious; if, on the contrary, in some given quantity they serve a useful purpose, and especially if they exercise a biogenic function, then it must be concluded that they are not essentially injurious even though they may become toxic by reason of what the author has elsewhere[1] termed an abuse of the quantitative relation.

EXPERIMENTAL OBSERVATIONS

The inimical action of soluble aluminum salts, even in relatively small amounts, would seem to be established by the following experimental observations:

Work of House and Gies. In 1905, House aud Gies,[2] studying the effects of soluble aluminum salts on the growth of lupin seedlings, announced that in concentrations greater than m/65,536* growth usually was markedly inhibited and in concentrations less than this and greater than m/1,048,-576 or m/2,097,152, growth was, as a rule, stimulated. On this evidence they concluded that such compounds, even when present in very minute proportions, are strong "protoplasmic poisons" and very toxic to growing plants.

Work of Hartwell and Pember. In 1918, Hartwell and Pember,[3] of the Rhode Island Experimental Station, reported that in studying acid soil conditions, they observed that liming exerted very little influence upon the growth of rye, yet the growth of barley was increased two to threefold. Investigating by the method of water culture, barley seedlings were not more susceptible than rye seedlings to injury by acidified nutrient solutions. This difference led the authors to investigate further in an effort to find out why barley was more susceptible than rye to acid soils and not to acidified water cultures. They concluded that the difference was due to the presence of soluble aluminum compounds in the acid soil which exerted a toxic effect upon barley but which did not, at least to anywhere near the same degree, exert a toxic effect upon rye. By the addition of phosphoric oxide or acid phosphate to the soil whereby the active aluminum was much decreased, even though the soil acidity was actually increased thereby, a remarkable growth of barley was obtained.

Work of Conner. In 1904, S. D. Conner, Chemist at the Agricultural Experiment Station, Purdue University, Lafayette, Indiana, reached the conclusion that the failure of corn to grow productively in such soil was due to the acidity rather than to the soluble aluminum salts present. Notwithstanding this conclusion, with the announcement by Hartwell and Pember that soluble aluminum salts were a causative factor of such unproductiveness of barley, he resumed his investigation of the subject and reversed his former opinion.[4,5] It appeared that in the water cultures of corn that constituted a part of the early investigation, he had failed to recognize that the nutrient solution employed contained phosphate and that both phosphate and silicate have the property of

* m = molecular weight in grams per liter. The concentration of aluminum in a m/65,536 AlCl₃ solution is 0.00004 per cent.

inactivating aluminum by forming insoluble compounds, thus throwing it out of solution. Hence, in the presence of phosphate there was no inhibition of growth attributable to aluminum. When, however, in the later work, a nutrient solution containing very much less of phosphate was employed, Conner found that in cultures with solutions of aluminum salts of the same hydrogen-ion concentration as the acid cultures without aluminum used for comparison, inhibition of growth was much greater; that is, that the acid salts of aluminum were more toxic than acid of the same hydrogen-ion concentration. The hydrogen-ion concentration of the acid soils was pH 4.2 to 4.8, at which acidity intensity the aluminum would precipitate as phosphate; while even at pH 6.0 it is not precipitated as the hydroxide. It was for these reasons that liming the soil failed to produce good crops while with the further addition of phosphate good crops were obtained. The work was extended with similar results to red beets, pop corn and buckwheat.

Studies of Osterhout. W. J. V. Osterhout[6] has studied the action of inorganic salts, including salts of aluminum, in connection with the subject of "injury, recovery and death in relation to conductivity and permeability." In his researches he measures by exact quantitative method the electrical conductivity of organisms, employing particularly the marine plant, Laminaria, and also the skin of the frog, and finds that whereas the electrical resistance is under normal conditions a certain normal degree, that with injury the resistance steadily falls, and if the condition producing the injury continues to act, the death point is eventually reached, at which point the resistance is only that of the surrounding solution. The changes in conductivity which occur under the influence of chemical substances run parallel to changes in the permeability of the protoplasm but depend not only upon the protoplasm, but also upon the cell wall and the cell sap.

If a series of observations of the resistance be expressed as ordinates and the time as abscissae, the resistance of a living tissue under normal conditions will be represented by a line that is practically parallel to the abscissae. On the

other hand, the injurious action of a given substance will
be represented by a curve. One group of salts which includes
the salts of monovalent metals causes a progressive loss of
resistance, as expressed by a simple downward curve. The
bivalent and trivalent cations are included in a second group
of salts which first produce a rise in resistance (stimulation
effect) followed by a fall which continues until the death
point is reached, which action will be represented by a curve
exhibiting an initial rise followed by a fall from the point
of maximum resistance to the point of resistance of the
containing solution, the death point. If the tissue is subjected to an exposure and withdrawn before it is too near the
death point, and placed in a corresponding environment not
containing the substance, there may be partial recovery,
a partial rise of resistance which will be indicated by an
upward bend of the curve, but not to the line of normal
resistance. The failure to reach the normal indicates permanent injury to the tissue by the particular acting substance.
It is equally true after exposure to the cation that first
raises the resistance, that recovery may not be to the line of
normal resistance but below that line, even if the discontinuance be made in the early part of the exposure when the
resistance is actually above normal, thus indicating that
there is injury even from the exposures to divalent and
trivalent cations that merely raise resistance.

Aluminum, in common with the divalent and trivalent
cations, first raises resistance and subsequently lowers it.
With solution of 0.02M* and if the action is for a sufficient
time the death point is ultimately reached.[7]

The different aluminum compounds behave in about the
same manner. As compared with iron, Osterhout stated that
the various salts of aluminum and iron sulphate and aluminum sulphate behaved about alike but that he would have
to look the matter up as to whether the iron compound was
the iron aluminum sulphate, as he believed, or was iron alum.
However, considering on the one hand this uncertainty and
on the other hand the facts that iron aluminum sulphate is

* M = molarity; i.e., the weight in grams per liter of the substance divided
by its molecular weight.

practically unknown and that iron ammonium sulphate is of relatively common occurrence, there is strong presumption that the substance actually used in his experiments was iron alum, that is, an alum in which the aluminum was replaced by iron. In this event, the observation that the aluminum compounds and the iron compound behaved about alike would lead to the conclusion that by this method iron and aluminum exhibit about the same degree of toxicity. In his earlier observations it is true, as he stated, that ferric sulphate produced less effect than did aluminum sulphate, but in those experiments the strength of the former solution was 0.0025M[6b] and the latter 0.01M. Any conclusion from available data that the method demonstrated greater toxicity of aluminum compounds than of iron compounds is therefore wholly unwarranted.

In connection with the action of aluminum in soils and in water cultures of seedlings, we have seen that any toxic effect is dependent upon the aluminum being in a soluble, ionizable form. In regard to his method, Osterhout expressed the opinion, as a general principle, that if the aluminum were in a non-ionized form it did not follow that it would be less toxic; it might be more so, and he cited the toxicity of organic iodine in support of this idea. Even if blood serum were the solution in which the aluminum salts were placed, he felt practically certain that it would make no difference. A little of the aluminum would combine with the protein but only a minute proportion of it. These opinions as to non-ionized aluminum expressed by Osterhout are not in harmony with the results of other actual observations and until supported by demonstrations can be given little weight.

Studies of Himebaugh. Lester C. Himebaugh[7] has conducted experiments to show the penetration of aluminum salts through a colloidal medium and to demonstrate their antiseptic action. For the colloidal medium he employed the bacterial medium standard-beef-extract-agar which was allowed to solidify in petri dishes in the center of which was placed in each case a solid, such as a glass stopper, about ⅞ in. in diameter, which was afterward removed, leaving a cavity or cup which was then sealed at the bottom by the addition of a few drops of the agar medium. The agar medium was seeded with staphylococcus aureus before the plate was poured. Into the cup of a plate thus prepared was placed ½ cc. of the particular solution of aluminum salt to be tested, after which the plates were incubated at 37°C. The strength of the solution of a particular aluminum salt tested was 0.05M. After forty-eight hours it was observed that immediately adjacent to the cup was a cloudy zone or ring; outside and adjacent to the cloudy zone was a clear zone; neither of these zones exhibited any growth of bacteria. Outside of the clear zone and comprising the remainder of the surface of the agar plate, bacterial growth was exhibited in the usual way. It was concluded that the aluminum salt penetrated through the agar medium of the plate producing the cloudy zone and exerted its antiseptic effect in both the cloudy and clear zones; the experiments thus proving, in Himebaugh's opinion, both the penetration of aluminum salts through a colloidal medium and the antiseptic action of such salts. Sulphuric acid of the same pH as the aluminum compounds failed to produce any such effect. Two per cent carbolic acid did.

Studies of E. E. Smith. E. E. Smith[7] called attention to the danger of applying the conclusions reached by the Himebaugh agar-cup method to conditions that exist in the body. He found that if the aluminum salts were mixed with blood and in this condition were introduced into the agar cup, neither the cloudy zone nor the antiseptic action was exhibited, the agar yielding bacterial growth to the very edge of the cup. Under body conditions, then, aluminum was inactivated by the albuminous fluids, such as blood, and neither affected adjacent tissue by the production of a cloudy zone nor under these conditions did the method give evidence of penetration through the colloidal tissues and fluids of the body.

Himebaugh,[7] using the mixture with blood suggested by Smith, thought there was an inhibition of bacterial growth for a very slight distance at the margin of the agar cup but not to the extent that occurred without the presence of blood. However, granting the correctness of his observation, penetration was limited to a distance that was almost negligible, in marked contrast to his results without the presence of blood. Hence, his observations did not disprove the general conclusion that albuminous fluids tend to inactivate the aluminum ion.

The test by admixture with blood likewise throws doubt as to the validity of the opinion expressed by Osterhout that aluminum in blood would show the toxic action exhibited by aqueous solutions, as measured by the diminished resistance of tissues in experiments conducted by his method.

Work of Stoklasa. In contrast to the opinions based upon the above various experiments to the effect that aluminum is essentially a protoplasm poison and is not of biogenic value are the results of experiments and conclusions drawn therefrom by Julius Stoklasa.[8] This investigator in 1911 published the view that aluminum is a protective substance and is indispensable to plants, a conclusion that seems fully established by the results of further extensive experimental studies by himself and collaborators. He recognizes aluminum, manganese and iron as biogenic elements, all taken by the plant cells from the soil according to their needs. Iron and manganese are distributed to the upper part of the plant, especially to the leaves and flowers, entering particularly into the formation of the chlorophyl structures and thus exercising their function, while aluminum is localized chiefly in the roots, rhizomes, root tubules and bulbs.

The action of the aluminum in saline combination is at least twofold. In gram-atomic* aluminum concentrations from .0001 to .0005 it acts as an activating agent upon germination, as was shown experimentally with the chloride in studies on germination which showed the following optimum increase above the controls, moistened with distilled water, of the solid matter of the rootlets after growth for ten days, during which time the test seeds were continuously moistened with distilled water containing the aluminum salt: Wheat, +11 per cent; pea, +14 per cent; barley, +31 per cent; peppergrass, +41 per cent. In the wheat cultures the alumi-

* The gram-atomic concentration of an element is the weight of the element in grams per liter, divided by its atomic weight.

num concentration was 0.0005; in the cultures of the pea, barley, and peppergrass, 0.0002.

Not only aluminum salts in weak concentrations thus give a well-marked impetus to germination, but this property is possessed by manganese to a lesser degree.

If aluminum chloride in gram-atomic aluminum concentration of 0.005 be employed, germination is depressed instead of augmented. The same is true if manganese chloride of this concentration be employed, the depression with manganese being as follows: Wheat, −23 per cent; barley, −33 per cent; pea, −34 per cent; peppergrass, −40 per cent. If in addition to the manganese chloride of gram-atomic manganese concentration of 0.005, aluminum chloride of gram-atomic aluminum concentration of 0.0002 in the case of wheat and barley be added, the depression due to manganese is lessened to −13 and −11 per cent respectively; while with pea, if the gram-atomic aluminum concentration be 0.0005, the depression due to manganese is lessened to −25 per cent; and with peppergrass, if the gram-atomic aluminum concentration be 0.001, the depression due to manganese is lessened to −9 per cent. It thus appears that aluminum in suitable concentration not only directly activates germination but during germination it exercises a detoxicating effect upon the inimical action of manganese. Since, as we shall see, this detoxicating effect is likewise exerted during plant growth upon iron as well as manganese, there is reason to believe similar studies with iron in plant germination would likewise show a detoxication of iron in this early stage of plant life.

In the studies on plant growth, the effect is measured by the dry matter in the whole plant, instead of the roots only, after growth in nutrient solution for seventy-two days. Upon plants growing in their natural habitat under conditions supplying medium quantities of moisture and saline matter (mesophytes), the results with aluminum and iron as sulphates are presented in the following table:

Gram-atomic concentration of aluminum and iron (as sulphates) in nutrient medium		Grams of dry substance in 10 plants				
Aluminum	Iron	Wheat	Rye	Oats	Barley	Buckwheat
0	0	46.21	80.34	59.05	66.70	20.53
.0005	0	48.33	80.04	61.27	68.19	20.84
0	.0005	45.09	75.33	52.98	66.87	21.01
.0005	.0005	59.42	89.74	69.92	74.15	24.38
.00075	0	50.51	81.66	62.00	67.16	21.97
0	.00075	41.83	72.99	56.04	63.02	19.08
.00075	.00075	56.57	88.83	70.12	75.48	28.29
.001	0	47.83	79.56	59.94	66.82	18.84
0	.001	39.52	68.75	50.02	62.43	16.51
.001	.001	49.77	83.22	63.15	60.10	24.09
.002	0	36.53	60.26	46.25	52.62	11.45
0	.002	30.19	51.05	43.83	50.94	9.30
.002	.002	37.68	60.18	50.72	54.87	14.76
.0005	.002	50.47	72.99	60.23	67.10	20.89

From these results it is seen that aluminum exerted relatively little activating effect on plant growth and that when a concentration of 0.002 was reached a toxic action began to be manifest, which may be expressed in percentage of the control growth as follows: Wheat −21 per cent; rye −25 per cent; oat −22 per cent; barley −24 per cent; buckwheat −44 per cent. Iron, on the other hand, not only failed to exhibit an activating effect in lesser concentrations than 0.002 but with wheat, rye and oats exerted a depressing effect even in dilutions of 0.0005, while with barley and buckwheat the injurious action appeared with concentrations of 0.00075. With a gram-atomic iron concentration of 0.002, the depression in the production of plant substance reached the following amounts: Wheat −35 per cent; rye −36 per

cent; oat -26 per cent; barley -24 per cent; buckwheat
-55 per cent. It appears, then, that iron in the same concentration is distinctly more toxic to plant growth than is
aluminum.

What, however, is of even greater interest in connection
with plant growth is the detoxicating effect of aluminum
upon both iron and manganese salts. When added to the
nutrient medium containing an iron concentration of 0.002,
so as to give in addition an aluminum concentration of 0.0005,
in four instances the depression was wholly overcome, and in
the other instance to a considerable extent as the following
percentages of production of plant substance, in comparison
with the controls, clearly show: Wheat, $+9$ per cent; rye,
-9 per cent; oats, $+2$ per cent; barley, $+1$ per cent; buckwheat, $+2$ per cent.

The results strikingly show that the small amount of
aluminum salts overcomes the toxic effect of the salts of iron.
Similar results indicate a like detoxicating action of aluminum salts on the toxic effect of manganese on plant growth.
To quote, "It is clearly indicated that the aluminum ions
exercise a considerable detoxicating effect." Again, "All
of our foregoing experiments have led to the conclusion that
neither iron nor manganese are able to exercise reciprocal
detoxication."

Plants that grow under conditions of minimum water
supply (xerophytes), including some varieties of rock brake,
Solomon's seal, Bohemian iris, feather grass, knotweed, corn
spurry, alyssium, gentian, rib grass naturally took up very
little aluminum and did not tolerate it well.

On the other hand, plants growing in their natural habitat
under conditions supplying unusually large quantities of
water (hydrophytes), including certain ferns (bladder,
flowering), grasses (reed, manna, cane), sedge and certain
flowers (lily of the valley, buttercup, marsh marigold, cranes-
bill, sage, colt's foot) absorbed and tolerated aluminum even
up to a gram-atomic concentration of 0.006 without being in
any way injured.

In reference to the mode of action of the salts of aluminum,
manganese and iron, Stoklasa establishes that there is

clearly a parallelism between the degree of physiological activity and the amount of dissociation, as indicated by the electrical conductivity of the solutions of various salts. In keeping with this, in salt mixtures where the electric conductivity, as indicating the dissociation, is less than the sum of the conductivities of the individual components, there is correspondingly a lesser physiological action. Considerations such as these indicate that in the first instance, the physiological action is dependent upon the ability of the metallic substances to enter the plant by osmosis. Again, substances which make up the plant cells, as hemicellulose, pectins, etc., absorb cations, such as the aluminum, manganese and iron ions and form with the same less soluble or less permeable combinations. The occurrence of these substances in or closely adjacent to the cell membranes constitutes a second factor influencing the passage of these cations into the cell protoplasm and hence correspondingly influences their physiological activity. How the catalytic influence of the individual cations diffused into the cell is of value in affecting germination and growth of the plant is not clear. It at present constitutes a phase in the plant activity that has not been unravelled and must remain unsolved till we have further knowledge of the processes of plant metabolism. We can only say that the physiological action of the cations is the resultant of all the physical and chemical conditions which occur in plant germination and growth.

In brief, the work of Stoklasa and his collaborators establishes:

That the aluminum ion, and also the ions of manganese and iron, have a biogenic function in plant life.

That aluminum is localized and functions chiefly in the roots and corresponding parts of plants; while manganese and iron are localized and function chiefly in the leaves and flowers.

That the source of each of these ions is by absorption from the surrounding soil.

That the function of aluminum ions is chiefly twofold:

1. They activate the process of germination.
2. They detoxicate the manganese and iron in absorption and hence render possible their absorption and supply to

the chlorophyl structures, without injury to plant germination and growth.

That in amounts exceeding these that are of value to plant germination and growth, the ions of each of these metals exert a toxic action, the toxic action of iron being greater than that of aluminum.

The work of House and Gies accords with the conclusions of Stoklasa in that each finds aluminum ions in smaller amounts stimulating and in larger amounts toxic. However, House and Gies limited their observations to plant growth, while Stoklasa finds the activating action is more especially to the process of germination, and, further, they did not study the reciprocal relation of aluminum with manganese and iron and so did not observe its detoxicating function. The beginning of toxic effect noted by House and Gies corresponds to 0.000015 gram-atomic aluminum concentration, while that of Stoklasa was about 0.0015. Since the observations of the former were on a single plant, the lupin, while those of the latter were upon a large number and great variety of plants, it cannot be concluded that the latter was in error. It is possible that the explanation is in the fact that the observations of House and Gies were made by growth in water while those of Stoklasa were made by growth in an aqueous nutrient solution.

There appears to be no conflict between Stoklasa's work and the conclusions as to aluminum in soils. In the latter instances, the amounts of aluminum present are relatively large. No attempt was made to investigate the presence or absence of a biogenic function of the aluminum ions.

Osterhout's alum solutions of 0.02M correspond to a gram-atomic aluminum concentration of 0.04. Solutions of such strength are frankly toxic to plant growth by both methods of investigation. It is to be expected that the study of more dilute solutions by the Osterhout method would have yielded results in harmony with the biogenic functions of the aluminum ions, as recognized by Stoklasa. The work of the latter investigator disproves any conclusions to be drawn from Osterhout's results, either that aluminum is essentially toxic or that it does not serve a useful and indeed necessary purpose in plant life. It also establishes that the action of aluminum on plant life is to be attributed to the aluminum ions and not to fixed aluminum.

CONCLUSIONS

The evidence presented on the relation of aluminum to plant life leads to the following conclusions:

It is found that the activity of aluminum in plant life is dependent upon its ionization, that is, it is the aluminum ion and not the fixed aluminum that exhibits activity in plant life. In certain dilute concentrations, up to gram-atomic aluminum concentration 0.001 under the experimental conditions, the aluminum ions activate the process of plant germination and by a reciprocal relation with iron and manganese ions exert a detoxicating effect upon these ions, whereby they may be absorbed without injury to plant life. In certain greater concentrations, the aluminum ions, in common with the trivalent cations generally, are capable of exerting a toxic effect. Such an effect is not due to the fact that aluminum is essentially a protoplasm poison, since from what precedes such is not the case but to an abuse of the quantitative relation. Comparing atomic concentrations, iron is toxic to plant life in greater dilution than is aluminum.

An interesting side light on the biogenic function of aluminum in plant life is its distribution as related to its functional activity. From the data already presented in Chapter II it appears that its occurrence in plants is by far the greatest in roots, in seeds, etc. That is to say, it is most abundant in those parts where it exercises its biogenic function. It is fair to believe that this is not a mere accident but a provision in nature to supply to these parts the chemical elements useful and essential for plant germination and the reciprocal detoxication of absorbed iron and manganese which elements are necessary to the constitution of the chlorophyl structures essential to plant life.

IX

THE ACTION OF SALTS OF ALUMINUM ON UNICELLULAR ANIMAL LIFE AND ON THE ISOLATED CELLS AND TISSUES OF HIGHER ANIMALS

Work of Heilbrunn. Lewis V. Heilbrunn, instructor in zoology, University of Michigan,[1,2] has studied the influence of soluble aluminum salts upon the fluidity of the protoplasm of sea urchin's eggs. He determines the fluidity by measuring the movement of the protoplasm granules under centrifugal force, the distinctness of the zones produced and the width of the hyaline zone being the index of the fluidity.

The natural habitat of the eggs is sea water. If aluminum salts were added to the water to the amount of ten parts (Al) per million, the protoplasm became coagulated and after a moderate exposure was killed. Smaller amounts, even down to one-half part per million, increased the fluidity in the interior of the egg while frequently the outer margin of the cells became brittle and fragile and broke readily. Fluidity did not permanently injure protoplasm but the investigator believed it was practically certain that when in that condition no vital activities could take place. Other metals produced the same effect, in the order Al > Ca > Mg > K. Na. NH$_4$. The effect of aluminum was about a thousand times as great as that of calcium. Cerium was many times more powerful than calcium and lead was between calcium and cerium. Heilbrunn stated he had never worked with iron salts and knew nothing of the relative action of iron and aluminum.

Work of Remington and of Himebaugh. The influence of aluminum salts upon the isolated red blood corpuscles has been studied by two observers.

Roe E. Remington, professor of biological chemistry, North Dakota Agricultural College, demonstrated[2] the injurious action of aluminum salts to beef and human corpuscles suspended in sugar-salt solution by the diminution of resistance to an alternating current and the production of hemolysis.

Electrical conductivity (resistance) of the suspension was measured by the use of an alternating current of such frequency that it could be compared with telephones. On one side of the system was a resistance of known value, on the other a bulb for the suspension, containing two metal plates with wires passing out of the bulb so as to be connected into the circuit. Between the bulb and resistance, in one direction was a fixed terminal from the source of the current; in the other direction a movable terminal which could be adjusted to the point where the current through the bulb and through the known resistance were equal, as indicated by the absence of the note through the telephones.

In making observations, the temperature was maintained constant, and the fluid in which the corpuscles were suspended was so compounded as to have the total concentration of dissolved substances and the electric conductivity the same as blood serum. With blood serum alone, with sugar-sodium chloride, with sugar-sodium sulphate, or with sugar-calcium chloride, the plotted line of resistance over a period of time was a straight line and nearly horizontal, showing only a slight loss of resistance; with sugar-aluminum sulphate, with sugar-aluminum chloride or with sugar-alums, (potassium, ammonium, sodium, or s.a.s.), there was at first a rapid increase of resistance followed by a gradual loss, the plotted line being a downward curve more striking at first and subsequently approaching a horizontal straight line. At the same time, in the aluminum cases and not in the others, the hemoglobin passed into the fluid (hemolysis). The strength of solutions employed corresponded to forty-five parts of aluminum per million. Equivalent amounts of hydrochloric acid and of sulphuric acid gave an initial increase of resistance, but the after part of the line was straight and not curved and there was no hemolysis. Lead salts gave the same results as aluminum, only more marked. There seemed to be some similarity of results with iron salts, but owing to the interference of the color from the iron salts, work with this element was not carried far enough to form an opinion. The conclusion reached was that the action of aluminum upon the red corpuscles caused substances to come out of the cells and at the same time the cells became more permeable to the passage of a current. Microscopically, they at first became crenated, followed by a secondary swelling, at which time the contents came out.

Work of Himebaugh. Lester C. Himebaugh, New York, studied[2] the action of aluminum salts on the washed guinea pig red blood corpuscles in the production of (a) agglutination, (b) primary hemolysis, and (c) a condition of the remaining corpuscles whereby they were abnormally resistant to secondary (saponin) hemolysis. When the blood corpuscles in suspension were acted upon by solutions of aluminum salts, agglutination of the corpuscles with s.a.s. was complete down to a dilution of 1:30,000; approximately 90 per cent with a dilution of 1:40,000; absent with a dilution of 1:50,000. There was hemolysis (primary) to the extent of 40 per cent in a dilution of 1:1000; 5 per cent in a dilution of 1:10,000; and none in a dilution of 1:15,000. On the residual corpuscles, saponin hemolysis (secondary) did not occur in dilution 1:1000 to 1:10,000; there was 10 per cent hemolysis in a dilution of 1:15,000; and complete hemolysis in a dilution of 1:30,000.

It was concluded that s.a.s. prevented the expected hemolysis (secondary) by its action upon the red blood corpuscles. Similar results were given with other aluminum salts. The effect due to sulphuric acidity was largely excluded.

Himebaugh[2] also studied the action of aluminum salts on isolated heart tissue. When pieces of the heart of an 8-day-old embryo chick were left for five minutes in a solution of .025 molarity of ammonium, potassium, or sodium alum, or of aluminum sulphate, and subsequently tested for the power of growing, they failed to grow. With .005 molarity some growth occurred; with .0025 molarity a good growth but not equal to the controls. This toxic action was not due to any considerable degree to the acid element, since sulphuric acid of the same hydrogen-ion concentration gave nearly the same growth as the controls.

CONCLUSIONS

In the above investigations of the action of the aluminum ion on unprotected unicellular and isolated animal cells and tissues, the gram-atomic aluminum concentrations (see foot-note, p. 176) employed were: Heilbrunn, .0004–.00002; Remington, 0.0016; Himebaugh, agglutination, greater than 0.002; primary hemolysis, greater than .007; tissue cells, .005 and up. It thus appears that Heilbrunn's results were obtained with lesser concentrations, Remington's with about the same concentration and Himebaugh's with greater concentrations than were found by Stoklasa to be injurious to plants. In Heilbrunn's experiments no consideration was given to the possible presence of aluminum in the sea water employed. However, such results, though interesting, are not strictly comparable, since in each instance the figures are only applicable to the particular conditions of the individual experiments.

Moreover, in some instances the investigators have made comparisons of the action of aluminum salts with the action under similar conditions of other metallic salts. Since the evidence points to the fact that in each instance it is the metallic ion that produces the effect, any accurate comparison calls for the employment not of solutions of the same atomic or molecular concentrations but of solutions of the same metallic ionic concentrations. No effort seems to have been made to regulate this factor, an omission that robs the comparative results of much of their value unless it should appear that in the considerable dilutions employed the metallic salts are entirely ionized and that undissociable combinations are not formed.

It is also noticed that there is an inclination to emphasize the comparison of the action of aluminum salts with the action of salts of lead, presumably because of the fact that lead is frankly poisonous to higher organisms, including man, the inference being that, if they act similarly on isolated and unprotected cells, they will act similarly on man. Such a conclusion is, of course, wholly unwarranted. It may be properly questioned whether or not they would so similarly act, but this is far from concluding that they do, since it entirely fails to give consideration to the factors of defence of higher organisms, including man. The latter are admittedly quite inadequate as a protection against lead, yet may be, and indeed, as will appear later, are entirely adequate as a protection against aluminum, precisely as they are against iron.

Unfortunately, none of the above investigators has thought it necessary or has been able adequately to compare the action of aluminum ions with those of iron. So far as such comparisons have been made, the action seemed similar. Now, iron is an element essential to the structure of the body, so it need not be questioned that there is a protective mechanism of the body that permits the ingestion and absorption of iron in amounts necessary for body use without injury. Moreover, there is a close relationship, as regards chemical behavior, between aluminum and iron, so that it is quite within reason that the same defensive mechanism, protective against inimical effects of iron, operates similarly in regard to aluminum.

Considering all the evidence at hand, what has been shown relative to the action of the so-called heavier metals is that the metallic ions are capable of producing injury to naked animal protoplasm. This conclusion, so far as there is any evidence,[3] is quite as applicable to iron, a natural constituent of the tissues of higher animals, as to the ions of aluminum and other metals. The applicability of this to the action of metallic ions, ingested in alimentation, upon the cells and tissues of higher animals, including man, is dependent upon the adequacy or inadequacy of the protective mechanism of such higher animals for the individual metallic elements. In the experiments recorded no evidence is adduced bearing upon this aspect as part of the subject, so essential to any conclusion as to the applicability of the results to man.

Further, no effort has been made in the study of the action of aluminum upon animal protoplasm to determine the effect of the undissociated as distinguished from the ionized element. Indeed, in all the above work, conditions have been established favorable to ionization and avoiding fixation, as though the latter might possibly, as indeed it seems certain it would, fail to show the unfavorable effects noted when ionized.

Until recently, few efforts have been made to determine whether aluminum exercises a biogenic function in relation to the life of animal cells and tissues. While there are no *a priori* grounds for anticipating such a function, the possibility must be recognized, especially in view of the exercise of such functions relative to plant germination and growth. Investigations suggestive of such a function in the albino rat have recently been reported (see p. 293).

X

THE SOLUBILITY IN THE GASTROINTESTINAL TRACT OF THE ALUMINUM COMPOUNDS OF BAKING POWDER RESIDUES

In the middle of the last century one of the reasons put forward as to why alum added to flour was injurious was that it formed an insoluble combination with the phosphate of flour, preventing its absorption and thus depriving the body of an essential constituent of this food. In the Norfolk Baking Powder litigation (p. 146) a similar view was presented in respect to the aluminum of baking powder residue. Presumably, it was inferred that if the hypothesis in regard to alum in flour were true, the same reasoning was applicable to the aluminum of baking powder residue.

Siem's studies (1886) on the effects of the subcutaneous injection of soluble aluminum salts may have stimulated interest in the question of the solubility and absorption of the aluminum of baking powder residue, so that if aluminum were soluble and absorbed it might be held responsible for producing the same injurious effects as had been found by Siem through its subcutaneous introduction into the body of animals, especially in the light of Mott's declaration that aluminum accumulated in the various organs, interfering with their proper functions.

EXPERIMENTS IN VITRO

In 1887, Pitkin (p. 49), working in vitro with canine gastric juice, calculated that approximately one-fifth to one-sixth of the total amount of the aluminum of baking powder residue in food was soluble in the human stomach; in 1888, Mallet (p. 51) using at body temperature 0.25 per cent hydrochloric acid-pepsin solution, to which was added aluminum hydroxide in one series and aluminum phosphate in another, concluded that from 47.3 to 61.8 per cent of the former and from 38.2 to 49.1 per cent of the latter passed into solution in two to two and one-half hours.

In recent years, the solubility of the aluminum of baking powder residues has been the subject of investigation by A. K. Balls.[1]

The bread used was made with 16 gm. of a straight (31 per cent) s.a.s. baking powder to 550 gm. of flour, 8 gm. of sugar and 5 gm. of salt, together with 400 cc. of water. In the solubility determinations, 5 gm. of bread were treated with 1 liter of the solvent. In the tables, column 1 was very well baked bread containing 5.84 per cent moisture; column 2 was well baked bread containing 7.09 per cent moisture; column 3 was poorly baked bread containing 16.09 per cent moisture; column 4 was poorly baked bread and an equal weight of lean meat. Time of each process was 24 hours.

TABLE LXXII

SOLUBILITY OF THE ALUMINUM RESIDUE OF BREAD MADE WITH STRAIGHT S.A.S.

Baking powder	Aluminum in solution							
	1		2		3		4	
Solvent	Mg.	Per cent	Mg.	Per cent	Mg.	Per cent	Mg.	Per cent
Distilled water..........	0.2	2.0	0.85	5	0.9	8		
0.2 per cent HCl........	10.2	86.0	10.6	93	9.1	85		
0.2 per cent HCl + 0.1 per cent pepsin..........	11.2	93.0	11.45	100	10.25	96	10.3	96
0.2 per cent HCl + 0.1 per cent pepsin followed by 0.5 per cent Na₂CO₃ + 0.2 per cent trypsin.................	1.25	11.0	0.75	7	1.1	10	0.85	8
0.2 per cent HCl, Al in dialysate.............	9.45	79.0						
0.2 per cent HCl + 0.1 per cent pepsin, Al in dialysate.................	9.9	83.0	.....	.....	10.7		7.4	69
Total Al, as AlPO₄......	12.0		11.4					

The figures give the milligrams of $AlPO_4$ obtained in the soluble residues from 2.5 gm. bread (500 cc. extract) and the per cent of the total aluminum of the sample which the quantity of aluminum of such residues represents. Only average figures are cited, Table LXXI.

One gm. of freshly precipitated aluminum phosphate was dissolved in a liter of 0.2 per cent HCl. The addition of 20 gm. Witte peptone produced no precipitate but, as appears (1) below, only about one-half the aluminum was then dializable through collodion.

TABLE LXXIII

SOLUBILITY OF FRESHLY PRECIPITATED ALUMINUM PHOSPHATE

Solvent	Aluminum Phosphate in Solution	
	Mg.	Per Cent
1. 1 liter 0.2 per cent HCl + 20 gm. peptone. 50 cc. used. $AlPO_4$ in dialysate.....................	24.3	49
2. 100 cc. of (1), neutralized and Na_2CO_3 added up to 0.5 per cent....................................	0.6	0.6
3. 100 cc. of (2) + 0.2 per cent sodium taurocholate..	0.8	0.8

From data such as these Balls concluded that during digestion "the gastric juice at that time will contain a high concentration of aluminum" with some absorption in the stomach; that in the duodenum before neutralization absorption may be rapid; "while in the lower portions of the intestines the amount of metal taken up by the body is probably negligible."

Assuming that the above observations are correct as to facts, there are certain criticisms of the conclusions. Thus 12 mg. as $AlPO_4$ per one-half liter corresponds to a concentration, as $AlCl_3$, of 0.0026 per cent. Even if this entire amount were in solution it is difficult to appreciate how it can be regarded as a "high concentration of aluminum." In determining solubility, Balls employed a proportion of bread to solvent that would yield a maximum percentage figure for solubility rather than represent the relation that there is reason to believe would exist in the gastric digestion of ingested baking powder bread. In his experiments the amount of solution in which it was soluble is relatively very large. It is not at all indicated or probable that if the relation of bread to solvent had corresponded to the amount that would exist in the stomach after eating an average amount of baking powder bread, say 100 gm., any such percentage of the aluminum would have been dialyzable or even have passed into solution at all. Therefore, since the assumption that increasing the amount of bread would have proportionately increased the amount of aluminum dissolved is unjustified, the conclusion that there has been demonstrated a high concentration of soluble and dialyzable aluminum is entirely gratuitous. The figures stand only for what they are, namely, the solubility in a liter of acid with and without pepsin exposed for a period of twenty-four hours in vitro of baking powder residue contained in 5 gm. of bread. Further, the assumption of absorption because of solution and dialysis through a collodion membrane ignores entirely the fact that absorption in vivo is a physiological process in which there is a participation of factors other than those concerned with the process in vitro. The secretion of mucus to combine with the mineral element and render it non-absorbable; the selective action of absorption, as exists for example, in the case of magnesium sulphate (which, though readily soluble and diffusible, is practically unabsorbed from the digestive tract, certainly to nowhere near what would be expected from its action in vitro), factors such as these are entirely ignored in Balls' physiological applications of his findings. In consequence of this his conclusions are wholly unwarranted. Whether there is absorption or not cannot be concluded from the limited information available from the observations under consideration.

In recent testimony,[2] Reginald Wright and Andrew Dingwall, chemists of the Royal Baking Powder Company, stated that in the case of bread made with a straight s.a.s. powder, they found a solubility of the aluminum residue of 70 per cent; in the case of a bread made with an s.a.s. phosphate powder, a solubility of 80 per cent. In the case of the latter bread, 65 per cent of the aluminum was dialyzable through collodion and through parchment. In all cases, the solvent was 0.2 per cent hydrochloric acid, both with and without an indefinite amount of added pepsin and diastase, exposed over night at room temperature. Some of the statements were contradictory and lacked clearness as to details.

Dr. Hal T. Beans, professor of chemistry, Columbia University, recently testified[2] as to the results of personal tests of the solubility of the aluminum reaction products of baking powder residue.

With mixtures of s.a.s. and sodium bicarbonate similar to straight powders, he concluded that the aluminum reaction product was insoluble in water but soluble in varying amounts from 42 per cent to roughly 50 per cent in .03N HCl, at 37°c. for thirty minutes.

With mixtures of s.a.s., calcium acid phosphate and sodium bicarbonate similar to combination powders, he found the aluminum reaction product partially soluble in the faintly alkaline solution resulting from treatment with water and completely soluble in .03N HCl at 37°c. for thirty minutes.

He concluded that the admixture of calcium phosphate with the constituents of a straight powder to form a combination powder had the effect of rendering a portion of the aluminum reaction product soluble in the faintly alkaline solution resulting from the treatment of this baking powder with water and rendering all instead of part of the aluminum reaction product soluble in .03N HCl.

In the course of his experiments he used s.a.s. and calcium acid phosphate with sodium bicarbonate in molecular proportions, and varied the amount, but never actually determined excess alkalinity and did not know whether the presence of an excess of sodium bicarbonate would render the aluminum reaction product more soluble.

EXPERIMENTS IN VIVO

We have already noted the failure of Patrick (p. 44) in 1879 to find soluble aluminum in the contents of the gastrointestinal tract of cats, during the active digestion of food with an alum baking powder.

Thirty years later, Prof. J. W. Mallet, in a personal statement to Dr. W. J. Gies,[3] dated August 27, 1909, reported that he had tested the solubility in gastric contents of aluminum compounds contained in bread made with a straight "alum baking powder," Bob-white brand, using 16 gm. to 550 gm. of flour. A test meal of the bread with water was taken in the case of each of 3 men on an empty stomach and the stomach contents removed after one hour and subjected to analysis. The following results were obtained:

TABLE LXXIV

Subject	A	B	C
Test meal			
Water, cc........................	175	175	325
Bread, gm........................	128	85	171
Aluminum from baking powder, as			
Al_2O_3, mg......................	107	71	143
Stomach contents withdrawn, cc.........	156	182	197
Aluminum in solution, as Al_2O_3, mg......	21	23	13
Per cent of total ingested............	19.6	32.4	9.1
Per cent of solution, as Al...........	0.007	0.0066	0.0035
as Al Cl_3........	0.035	0.033	0.017

It was concluded that no inconsiderable part of the aluminum was dissolved in the stomach.

About 1916, P. B. Hawk and C. A. Smith[2] investigated the solubility of the aluminum residue in food prepared with a straight (30.75 per cent s.a.s.) baking powder by determining the amount in solution in the mixed stomach contents removed at the expiration of thirty, sixty and ninety minutes after the ingestion of a test meal of such food. Muffins were made after the recipe:

Flour.. 2 cups
Bran... ½ cup
Butter... ¼ cup
Sugar.. ¼ cup
Milk... 1 cup
Egg.. 1
Baking powder................................... 20 gm.

TABLE LXXV

Subject	F	S
Test meal		
Water, cc..	500	500
Muffins, gm.	135	135
Aluminum from baking powder, as Al_2O_3, mg.	325	325
Stomach contents withdrawn, cc.	250	290
Aluminum in solution, as Al_2O_3 mg.	23.9	40.3
Per cent of total ingested.	7.35	12.4
Per cent of solution, as Al.	0.005	0.007
as Al Cl$_3$.	0.025	0.036

Recently, Victor C. Myers and J. A. Killian[2] reported investigations on the solubility of aluminum of the residue of food prepared with s.a.s. (straight and combination) baking powders, in both the gastric and the duodenal contents (see Tables LXXVI–LXXIX), using for subjects hospital patients not acutely ill, and withdrawing the stomach contents one hour after eating a known quantity of the test bread; and, in the study of the duodenal contents, beginning one-half hour after the ingestion of the food and continuing the withdrawal for one and one-fourth hours, the tube being introduced 85 cm. to 100 cm. from the mouth.

In the gastric contents, removed one hour after eating, they found, with bread not containing baking powder residue, aluminum present in solution, per 100 cc., in the amounts as Al_2O_3, 0.20 mg. and 0.15 mg.

No corrections for blanks are made in these various analyses, so it is questionable, in the first case of Table LXXIX whether any aluminum from the baking powder residue was present in solution.

To determine whether the very much smaller amounts of aluminum in solution in the duodenal contents was due to dilution of the gastric contents by the duodenal juices or to precipitation of the aluminum in solution: To 50 cc. gastric contents containing in solution 2.56 mg. as Al_2O_3, per 100 cc., from food prepared with a straight s.a.s. powder, 50 cc. normal fresh duodenal juice, alkaline in reaction, was added

TABLE LXXVI

NO. 1 COMBINATION POWDER CONTAINING 2.85 PER CENT AL$_2$O$_3$

Subject	L. H.	L. K.	H.	T. J.	J. A.	M. P.
Test meal						
Water, cc.............	400	400	400	400	400	400
Bread, gm.............	92	89	72	62.5	102	95
B.P., gm. per 100 gm....	6.6*	6.6*	1.6	1.6	1.6	1.6
Aluminum from B.P., as Al$_2$O$_3$, mg...........	172.9	167.3	32.5	28.4	46.3	42.9
**Stomach contents withdrawn, cc...........	180	250	195	350	284	180
Aluminum in solution, as Al$_2$O$_3$, mg.......	27.7	15.25	5.4	0.77	4.23	4.01
Per cent of total ingested...........	16.0	9.1	16.7	2.7	9.1	9.3
Per cent of solution as Al................	0.0082	0.0033	0.0015	0.00012	0.00079	0.0012
As AlCl$_3$..........	0.040	0.016	0.0073	0.00057	0.0039	0.0058
Average† as AlCl$_3$	....	0.028	....	....	....	0.0044
Acidity pH.............	2.18	2.35	1.85	1.38	1.38	2.38
Total, per cent, as HCl.............	0.262	0.149	0.197	0.197	0.193	0.153
Free HCl, per cent...	0.058	0.022	0.058	0.109	0.102	0.025

* Four times the amount indicated on the label.
† For 6.6 gm. and 16 gm. baking powder breads.

TABLE LXXVII

NO. 2 COMBINATION POWDER CONTAINING 4.19 PER CENT AL$_2$O$_3$

Subject	J. R.	M. B.	M. H.	M. K.
Test meal				
Water, cc......................	400	400	400	400
Bread, gm......................	104	101	68.4	92.7
B.P., gm. per 100 gm..............	2.73	2.73	2.73	2.73
Aluminum from B.P., as Al$_2$O$_3$, mg.	119.2	115.4	77.6	106.4
Stomach contents withdrawn, cc....	280	140	87	190
Aluminum in solution, as Al$_2$O$_3$, mg.........................	23.7	1.778	3.22	3.04
Per cent of total ingested......	19.9	1.5	4.1	2.9
Per cent of solution, as Al......	0.0045	0.00067	0.00196	0.00085
as AlCl$_3$.....	0.022	0.0033	0.0097	0.0042
Average as AlCl$_3$	....	....	....	0.0098
Acidity, pH....................	2.18	3.0	2.08	3.0
Total, per cent, as HCl........	0.255	0.182	0.153	0.124
Free HCl, per cent............	0.080	0.00	0.058	0.00

TABLE LXXVIII

NO. 3 STRAIGHT S.A.S. POWDER CONTAINING 6.95 PER CENT AL$_2$O$_3$

Subject	L. H.	I. S.	E. R.	G. S.	L. G.	C. D.	W. K.	M. P.
Test meal								
Water, cc........	400	400	400	400	400	400	400	400
Bread, gm........	85	80	86	99	92	105	91	91
B.P., gm. per 100 gm............	1.42*	1.42	1.42	1.42	1.42	1.80	1.90	1.20
Aluminum from B.P., as Al$_2$O$_3$, mg............	84.0	79.2	85.0	97.4	89.7	131.7	120.1	75.8
Stomach contents withdrawn, cc...	200	345	214	280	200	175	210	200
Aluminum in solution, as Al$_2$O$_3$, mg	2.66	16.9	8.77	3.07	16.3	5.28	20.8	5.3
Per cent of total ingested......	3.2	21.3	10.3	3.2	18.2	4.0	17.3	7.0
Per cent of solution, as Al..	0.0007	0.0026	0.0022	0.00058	0.0043	0.0019	0.0053	0.0014
As AlCl$_3$......	0.035	0.013	0.011	0.0029	0.021	0.0091	0.026	0.0069
Average as AlCl$_3$....	......	......	......	......	......	......	......	0.0156
Per cent of total Al of contents withdrawn...	25.0	32.6	19.9	......	39.4	14	53.4	......
Acidity Total, per cent, as HCl......	......	0.197	......	......	0.142	0.182	0.235	......
Free HCl, per cent........	......	0.00	......	......	0.066	0.077	0.096	......

* Quantity indicated on the label.

and after twenty-four hours at 37°c. the aluminum in solution was determined. Calculated to the original gastric contents, the amount found in solution was 0.82 mg. per 100 cc., indicating that the lesser amount in the duodenal contents was due to its precipitation from the gastric contents by the duodenal juice.

From the data presented, the conclusion is indicated that following the ingestion of straight s.a.s. baking powder food a certain amount of the aluminum compound of the baking powder residue may, and probably ordinarily does, pass into solution in the gastric contents. The maximum concentration found was 0.0082 per cent figured as Al, or 0.04 per cent figured as aluminum chloride. The usual concentration was

TABLE LXXIX

NO. 4 STRAIGHT S.A.S. POWDER CONTAINING 6.95 PER CENT AL$_2$O$_3$

Subject	G. R.	J. N.	G. S.
Test meal			
Water, cc................................	400	400	400
Bread, gm..............................	94	122	82
B.P., gm. per 100 gm..................	1.77	1.77	1.77
Aluminum from B.P., as Al$_2$O$_3$, mg......	116	149	101
Intestinal contents withdrawn during			
seventy-five minutes, cc............	65	60	80
Tube down, cm.....................	85	82.5	100
Aluminum in solution, as Al$_2$O$_3$, mg...	0.167	1.17	1.25
Per cent of total ingested..........	0.14	0.79	1.23
Per cent of solution, as Al.........	0.00014	0.00103	0.00083
as AlCl$_3$........	0.00067	0.00509	0.00409
Average as AlCl$_3$.................	....	....	0.00328
Per cent of total Al of contents with-			
drawn..............................	....	....	11.0

very much less. In the limited number of observations of intestinal contents removed by tube introduced through the pyloric orifice, the figures for solubility yielded a maximum concentration of 0.001 per cent, figured as aluminum, and 0.005 per cent figured as aluminum chloride.

THE ACTION IN THE ALIMENTARY TRACT OF FOOD PREPARED WITH S.A.S. BAKING POWDERS

THE TASTE

The action in the alimentary tract of food prepared with s.a.s. baking powders, either straight or combined, is an essential part of the inquiry as to whether such food is or may be injurious to health.

Since the enzyme activity occurs chiefly while the food is in the fundus of the stomach, the immediate relation of the food to the mouth has to do chiefly with the taste of the food and the secretion of saliva and its ptyalin together with the mechanical processes of mastication and deglutition. The statement has been made that food prepared with s.a.s. powders has a bitter taste. To demonstrate this, Dr. Ula M. Dow,[1] a teacher of Foods and Household Management, made biscuits after the following recipe:

Flour	1	cup
Butter	1	tablespoonful
Salt	½	tablespoonful
Milk	⅜	cup
Baking Powder		

Three sets of biscuits were made, differing only in the amounts, as level teaspoonfuls, and kinds of baking powder:

Set 1	No. 1	3 teaspoonfuls s.a.s. powder	
	2	2 teaspoonfuls s.a.s. powder	
Set 2	3	3 teaspoonfuls tartrate powder	
	4	2 teaspoonfuls tartrate powder	
Set 3	5	2 teaspoonfuls s.a.s. powder	
	6	2 teaspoonfuls tartrate powder.	

These biscuits were submitted to the taste of a number of individuals who variously reported that they found the s.a.s. biscuits bitter, salty, etc. In her own home she had used "alum sulphate" baking powder and had not found it "noticeably" unpalatable. In the above test, she used only a single can of powder, given to her for the purpose of the test by a party in litigation. It did not appear whether the particular can used was representative of a straight s.a.s. powder or whether by accident or incompetent manufacture it contained an excess of either s.a.s. or sodium bicarbonate.

Benjamin H. Jacobs,[1] Chemist, formerly of the cereal laboratory of the Bureau of Chemistry, U. S. Department of Agriculture, duplicated the biscuit-tasting experiments of Miss Dow, using the same recipe and the same brands of tartrate and s.a.s. baking powders and submitting them to four experimental tasters. He took the added precaution of allowing only one taster in the room at a time, since a gesture or a grimace from one person might inadvertently influence the opinion of another. They reported (labeled as in experiments of Miss Dow):

Taster

No. i. Biscuit Nos. 1 and 5, bitter

No. ii. Biscuit No. 2, slightly bitter; No. 5, bitter; Nos. 1, 2, 3, 4, and 5, salty; No. 6, slightly more salty

No. iii. Biscuit Nos. 1 and 2 tasted like baking powder; Nos. 3 and 4, flat; Nos. 5 and 6, normal.

No. iv. All biscuits tasted of too much baking powder

It was Mr. Jacobs' opinion that the biscuits contained twice and three times too much baking powder, four times too much salt and too much shortening. The butter and milk impart very distinctive flavor to foods, tending to mask other flavors. The idea of the excess quantities, he supposed, was to accentuate the baking powder flavor which was in part defeated by the quantities of salt and the character of the shortening. A bland shortening should be used. In tests of this sort it is the custom to eliminate as much as possible other contributing factors. For these reasons he did not consider either Miss Dow's or his own experiments competent to decide the point as to what the baking powders had contributed to the flavor of the biscuits.

Mr. Jacobs reported further experiments with various baking powders after the recipe:

Flour.......................... 226   gm. (2 cups)
Salt........................... 4½ gm. (¾ teaspoonful)
Shortening (Crisco)............ 28   gm. (2 tablespoonfuls)
Water.......................... 180   cc. (¾ cup)

The experiments were made in duplicate, employing two varieties of flour. The baking powders were:

No. 7   Tartrate, 2 level teaspoonfuls
No. 8   Straight s.a.s., 2 level teaspoonfuls
No. 9   s.a.s. phosphate, 2 level teaspoonfuls
No. 10   s.a.s. phosphate, 2 level teaspoonfuls.

Nos. 7 and 8 were the powders of the previous tests.

Each of the four tasters reported that these biscuits all had a normal taste; there was nothing abnormal, e.g., nothing bitter, salty or metallic, about them.

The taste of baking powder biscuits lends itself readily to observation of people generally, although individuals vary much in the acuteness of their taste appreciation. The writer has rarely if ever detected in any biscuits a taste that could be described as bitter, even when a foreign taste has been observed by others. Attention to the judgment of others has led to the opinion that what sometimes has been described as a bitter taste has been limited to food prepared with a straight s.a.s. powder probably used in excess of the quantity indicated and accompanied by a relatively large quantity of salt. Particular attention to food prepared with powders known to be made with ingredients accurately balanced, more especially the s.a.s. calcium acid phosphate powders (and such powders constitute the bulk of the s.a.s. baking powders of the present day), has not disclosed in the writer's experience and observation any foreign taste, bitter or otherwise, that can be ascribed to the baking powder ingredient.

At all events, if by chance food prepared with a particular baking powder does have a slight foreign taste, that fact in itself does not constitute an injurious effect on health. Such a circumstance will naturally be corrected by avoiding the use of excessive amounts of salt and baking powder.

The writer knows of no data determinative of any action of food prepared with an s.a.s. baking powder upon salivary secretion, mastication or deglutition. Hence, so far as it is known or there is any reason to believe, such food is without unfavorable effect upon these functions.

THE EFFECTS OF ALUMINUM COMPOUNDS ON ENZYMIC ACTIVITIES IN VITRO

Studies of the action of aluminum compounds on enzyme activities in most instances have been directed to the digestive enzymes.

In the examination of this subject, it should be recognized that the quantitative measurement of enzymic activity in vitro, while of biological interest and hence a proper consideration from that viewpoint, is by no means a measurement of the digestive process in vivo. Among other possible reasons for this are that in vitro there is (a) no removal of products of digestion which themselves inhibit enzyme activities; (b) no motion, similar to the continuous movements of the alimentary tract, which promote the digestive process; and (c) no physiological adaptation corresponding to the secretory process in vivo to regulate the supply of enzyme according to the amount required. Hence, digestion in vitro is not a measure of digestion in vivo.

In view of the more exact methods of modern investigations, the earlier pioneer work (see West-Knights, p. 44, Mott, p. 46, Pitkin, p. 49) is merely of historic interest. Ruttan (p. 50) compared the digestive activity of the substances in the residues from tartrate, phosphate and "alum" powders and concluded that in general there was retardation with tartrate, phosphate and aluminum residues in the order named, but that at least with minimum amounts of enzyme the products of digestion soon become more powerful retarding agents than were the constituents from the powders. Calcium sulphate retarded the amylolytic action of saliva more than did the tartrate, aluminum or phosphate compounds; and it was its presence in phosphate that, he believed, made the retarding action of phosphate greater than that of the aluminum residue. He might well have made a similar allowance for the retarding action of the sulphates from aluminum sulphate decomposition. He concluded: "There is nothing in common between the specific destructive action of alum itself and the semi-mechanical, retarding influence of these products of its decomposition. A quantity of alum that would entirely prohibit ferment action will, when converted into its equivalent alkaline sulphate and aluminum phosphate, delay digestion only from 20 to 30 per cent." He worked with the bread crumb reduced to dried powder, since with fresh bread the samples could not be reduced to the same state of mechanical division, thus rendering comparisons between different series of experiments useless. Such a difficulty well illustrates the influence of relatively immaterial factors upon digestion, as measured by this method, and indicates the limitations that necessarily must be placed upon experiments in vitro in their application to digestive processes in vivo. These limitations he failed to evaluate in his general conclusions that the moderate retarding actions observed by him would be prejudicial to health.

Hehner (p. 52) employed methods that were not the accepted methods of his time and are even less acceptable today. His results reflected his preformed opinions, for which reason, apparently, to his mind they justified his conclusions.

The methods employed and the results obtained by Bigelow and Hamilton (p. 55) are much more worthy of consideration. Apparently, realizing the wide gap between digestion in vitro and digestion in vivo, they attempted to lessen that gap by studying the effect on the combined actions of the proteolytic enzymes of the gastric and pancreatic juices. They studied the influence of the aluminum compounds on the enzymes when such compounds were added to the flour before it was made into bread with yeast. Their results show inhibition relatively slight by aluminum hydroxide and 10 to 12 per cent greater with aluminum phosphate, and appear to be trustworthy under their experimental conditions.

In the relatively simple experiment reported by E. E. Smith (p. 63), using the Sjollema modification of the Stutzen method, the digestion of the protein of a bread made with a "straight" powder was the same as the digestion of a control bread made after the Liebig formula, showing under these experimental conditions no retardation.

In 1904, Victor C. Vaughan,[2] in an address before the International Pure Food Congress, reported the action of gastric juice to which aluminum hydroxide had been added upon coagulated egg white in capillary glass tubes, the results indicating retardation, as compared with controls. He also reported comparative digestion in artificial gastric juice of bread made with tartrate and with "alum" powders, finding greater retardation with the latter. The number of observations reported were limited and they have never been published in scientific literature. In so far as the methods were described, they do not justify quantitative conclusions.

John H. Long, as a member of the Referee Board, conducted comparative quantitative experiments (p. 111), concluding that the aluminum residues in biscuits, whether made with a straight or a combination powder, did not "exert any action which seriously interferes with digestion."

Frederick S. Hammett has testified[1] regarding comparative digestion experiments in vitro, with biscuits made with a straight and with a combination s.a.s. powder and with a sodium acid sulphate powder as a control. In terms of the control, biscuits containing 6 gm. of powder per 240 gm. of flour gave with pancreatic lipase, fat digestion to the amount of 67.7 per cent for the straight and 64.8 per cent for the combination powder; biscuits containing 17.2 gm. of powder per 240 gm. of flour gave fat digestion of 63.1 per cent for the straight and 50 per cent for the combination powder. Biscuits containing 6 gm. of powder per 240 gm. of flour gave with pancreatic trypsin and enterokinase from the duodenal mucosa, protein digestion to the amount of 72.7 per cent for the straight and 78.8 per cent for the combination powder; those containing 17.2 gm. of powder per 240 gm. of flour gave 70 per cent for the straight and 86 per cent for the combination powder. In starch digestion, both with saliva and with pancreatic amylase, there was practically no difference between the straight, the combination and the control biscuits. The biscuits used throughout the experiments were made under the direction of an employee of a tartrate baking powder company and the recipe was known to Dr. Hammett only as reported to him. The slight excess alkalinity of the powders used was neutralized, so that the results do not represent the action of the powders as manufactured. The methods employed were not described. Combination digestive tests, as employed by Bigelow and Hamilton (p. 55) were not employed, on which account the results of the two researches are not comparable. It may be assumed that each is substantially correct for the experimental conditions but, as has been previously explained, neither series of experiments is a true measure of digestibility in the body. Even on the assumption that the fat of the biscuits was as much as 10 per cent of the solids, the amount of fat that would escape digestion because of retarded lipase action would in absolute amount be quite small. If the biscuits were eaten with butter, the larger amount of fat in the digested mixture would in all probability show not only a small absolute amount undigested under the conditions of the experiment, but also a small relative amount. Further, there are no indications that under physiological conditions of digestion even such small absolute and relative amounts would not be digested to the same extent as in the controls; indeed, physiological tests show this to be the case (see pp. 67, 83, 97, 105).

THE DIGESTION IN VIVO OF FOOD PREPARED WITH S.A.S. BAKING POWDER

The action of the individual enzymes in the digestive tracts of higher animals and man does not lend itself to direct determination owing to the inherent difficulties presented by the structure and functioning of the digestive system. However, very definite information on the subject is obtained by the broader determination of the utilization of a particular food in question. Such utilization depends upon and thus measures the physiological processes of secretion, including salivary, gastric, pancreatic and intestinal, the digestion of such food under the enzymic action of the various digestive juices and the absorption of the products of such digestion.

The procedure consists of feeding to the individual for a definite number of days a diet of known composition, which may include a relatively large proportion of the food in question (in this instance food prepared with s.a.s. baking powder), and collecting and analyzing the stools, accurately marked and separated to correspond to the diet of the particular feeding. The amount of the nutrients fed, less the amount of nutrients in the stools, represents the amount of the diet utilized by the body during the period of the test diet. In the stools there is a relatively small amount of admixture of the secretions of the alimentary tract because of which fact the term "utilization" does not quite accurately express the result obtained. Hence, the term "availability" is employed, which is understood to express the facts actually determined. Either prior to or following the diet of the food in question, an exactly similar observation may be made of a diet in every other way the same excepting that the food in question is replaced by a control food of the same composition of nutrients as the test food. The availability of the control diet serves as a comparison for the diet containing the food under investigation. Obviously, if the food in question interferes with the secretion of digestive juices, with the enzymic action of the digestive juices or with the absorption of the products of such digestive action, either it will be compensated for or it will be revealed by a difference between the availability of the diet containing the food in question and the availability of the control diet. On the other hand, if there is no such difference in availability, it follows that the particular factor entering into the diet in question has not interfered with the utilization of such food. Determination of the availability of a food shows the extent of the processes of secretion, digestion and absorption as a whole; it does not show the individual stages.

Determinations of the availability of food prepared with s.a.s. baking powder were first made by E.E. Smith (pp. 67, 77), using a diet nearly two-thirds of which consisted of food prepared with a straight s.a.s. baking powder. The availability of the fats, proteins and carbohydrates was the same as that of a control diet, indicating that all the nutrients of the baking powder food were fully utilized by the body. Further observations of a similar nature were made on two subjects (p. 83) by comparison of the availability of s.a.s. baking powder food at the beginning and at the end of a period of four and a half months. During this time the subjects had continuously partaken of a diet containing a large proportion of cereal foods, all raised with a straight s.a.s. baking powder, so that the total amount of such food ingested was much in excess of what would ordinarily be eaten. The availability showed the same utilization of the food at the end as at the beginning of the period. There was no loss of ability to utilize such food fully and normally.

In the elaborate investigations of the Referee Board of Consulting Scientific Experts of the U. S. Department of Agriculture, data on the utilization of diets were obtained. Professor Chittenden, studying baking powder foods, reported (p. 97) some, but not marked, loss of nitrogen utilization, especially when sulphates were present with their laxative action, while fat utilization showed no variations to suggest any influence of the aluminum compounds ingested. Likewise Professor Taylor, studying (p. 124) the action of large amounts of s.a.s. and its products of reaction, found nitrogen utilization lower only when laxative action was produced by excessive amounts of sodium sulphate. Also large quantities of the products of reaction increased slightly the fat of the stools.

The evidence, then, is that ordinarily there is no lessened utilization in the alimentation of s.a.s. baking powder food by reason of the presence in such food of reaction products of the baking powder ingredients, but that utilization may be slightly lessened by the ingestion of amounts so large as to produce laxative action.

THE INFLUENCE OF S.A.S. BAKING POWDER FOOD ON THE PROCESSES OF ALIMENTATION

Aside from the digestibility and utilization, there is to be considered the possible effect of the food in question upon other processes of alimentation. Certain direct studies have been made of the influence upon gastric secretion as indicated by the composition of the stomach contents removed after a test meal of the food in question. Such comparative examinations by E. E. Smith on two young men (p. 62) showed no unfavorable effects, the composition of the gastric contents being practically identical, whether the bread of which the test meal consisted was with or without the baking powder residue. Similar examinations in the cases of the two young men at the beginning and at the end of a four and one-half months' dietary period (p. 82) showed some increase in gastric acidity at the end as compared with the findings at the beginning of the feeding period. No significance was attached to this, since the increase was well within the limits of physiological variation. However, recently it has been attempted to interpret this finding as evidence of "an irritant effect on the gastric mucosa."[3] That such an interpretation is not justified and that the slight increase is a physiological variation seems established, at least in the case of subject R, by the fact that this individual happened to be the same subject who was employed in the earlier tests described (p. 62) and that on that occasion, by the same test of gastric function, the acidity of the gastric contents was almost identical with the findings at the end of the experimental period, some two years later. In the case of this individual, the real variation appears to be not the apparent increased acidity at the end of the experimental period, but a slightly lower acidity at the beginning of the experimental period. This illustrates the possibility of error of interpretation due to ignoring reasonable physiological variations.

P. B. Hawk and C. A. Smith recently testified[1] relative to the results of their study of the influence of food prepared with s.a.s. baking powder on gastric secretion. Employing a test meal consisting of 35 gm. of dry, well-toasted bread and 250 cc. of distilled water, samples of the stomach contents were removed by the fractional method at fifteen-minute intervals until the stomach had emptied itself. Its contents were analyzed for total acidity, for total and free hydrochloric acid and for pepsin by the Mette method. Such a test was made on 2 subjects at the beginning, the middle and end of a six weeks' period during which they were on a special diet. This diet included a large proportion of food prepared with a straight s.a.s. baking powder. In these observations the ability of the stomach to secrete hydrochloric acid and pepsin, as influenced by the baking powder food diet, was studied. They conclude: "The ingestion of alumized bread during a period of six weeks had no marked influence on the secretion of gastric hydrochloric acid, but did have apparently a stimulatory action upon the secretion of pepsin." Details of the results are not given. If the stimulatory action upon the secretion of pepsin should be confirmed, the possibility suggests itself that on a diet in which a particular food makes up a large proportion, there is an adaptation of the amount of pepsin to meet the requirement of that particular diet. In such an event, the activity of the secretory process of the stomach is in no sense to be regarded as a pathological process but merely a physiological adaptation. That in the case of two individuals there appears to be more than the usual amount of pepsin secreted when on a diet in which the s.a.s. reaction products are present in unusual amounts, does not, of course, establish the fact that such is regularly the case, even though the findings be in every way exact. The fact that P. B. Hawk and C. A. Smith do not find an increase of acid secretion following a period of the baking powder food diet is confirmation of the conclusion of E. E. Smith (p. 82) that the findings in the two subjects of his experimental studies on a similar diet are to be regarded as physiological variations of particular individuals rather than an indication of any untoward effect.

Another phase of alimentation that must be considered is the possibility of delayed digestion resulting in absorption without changes in the amounts apparently utilized by the body. An objection to such delay has been suggested, namely, that an increase of fermentation and putrefaction in the intestines of the ingested foods would occur. In the experimental studies by E. E. Smith above mentioned, observations upon this point were made by exact determinations of such products eliminated by the kidneys after absorption from the bowels (pp. 83, 84). It was found that the elimination in the urine of indican and of combined sulphates was not increased by the diet in question, being if anything slightly less. In the more elaborate studies of the Referee Board, Chittenden found no significant changes in the fecal flora and no increase in the bacterial decomposition products in the feces, namely, phenols, indol, skatol and hydrogen sulphide; nor of the phenols or oxyacids eliminated in the urine. Likewise, the data obtained in the studies by Long (pp. 100, 105) were not indicative of any gastrointestinal toxemia from increased bacterial decomposition of the food. Evidence, therefore, is entirely lacking of increased bacterial decomposition with the suggestion therefrom of delayed digestion and absorption in the alimentation of such baking powder food.

In addition to the above exact data relative to the processes of alimentation, there are to be considered the clinical or symptomatic evidences of disturbances of the digestive processes. It is not necessary to cite data as to the production of various symptoms, such as astringent taste, gastric distress and bowel disturbances, including both constipation and diarrhea, that have from time to time been noted to result from the ingestion of considerable amounts and concentrated solutions of alum. It may well be and need not be questioned that alum is capable of producing such symptoms. So much emphasis seems to have been put upon these manifestations, in connection with the warfare against the addition of alum as an adulterant of flour, that, even as far back as the middle of the last century, Tardieu (p. 43) was led to pronounce his dictum, relative to salines, including alum: "The concentration is of more importance than the dose;" a generalization that is as applicable today as it was then. The necessity for the repeated pronouncement of Tardieu's dictum seems still to exist.

As a result of experiments with dogs, Mott reported (p. 46) the occurrence of vomiting after the feeding at one time of 8 biscuits containing $3\frac{1}{3}$ and $7\frac{1}{3}$ teaspoonfuls of a straight alum baking powder. The actual amounts at one feeding may have varied from 15 gm. to 100 gm. of baking powder according to his conception of a teaspoonful.

·This would be the equivalent of from 12 gm. to 80 gm. of Glauber's salts, of which the u.s.p. dosage for a human adult is only 15 gm. Dogs are notably prone to vomit; it ordinarily requires very little urge to promote this occurrence. It is not surprising, therefore, that some of his dogs vomited, and this without any consideration of the aluminum residue. An interpretation of the results of experiments of this kind would not seem to justify any conclusion whatsoever relative to the ingestion of baking powder food by man. In all probability common table salt in equivalent dosage would have produced the same results. His feeding to dogs of aluminum phosphate and aluminum hydroxide in ⅛ oz. to ⅜ oz. doses, even though accompanied by meat, is equally without bearing upon the question we are considering.

Mallet's observations (p. 51) of the experience of oppressive sensations following the ingestion of aluminum hydroxide and aluminum phosphate, eliminating the factor of suggestion which it is always difficult to evaluate, may be accepted as competent observations of the symptomatic effects on a single individual of these substances in the dose and manner ingested. However, even his relatively cautious conclusion that "the residues which the baking powders leave in bread cannot be viewed as harmless but must be ranked as objectionable," is one which his observations did not justify. Other articles of food, some of which are admittedly essential to life and are taken as a part of the food into whose composition they enter, in the same relative dose and under similar conditions of administration as employed by Professor Mallet in his observations, would in all probability produce equal or even more oppressive sensations than those that he experienced. No such sensations are experienced in the everyday ingestion of the baking powder residues in food, nor does most careful experimental study of baking powder food ingestion reveal digestive disturbances to account for his experiences. For these only one explanation would seem feasible, namely, that the conditions and results of his tests were not the conditions and results of normal baking powder food ingestion. The same criticism applies to Hehner's observations (p. 52) on the administration of baking powder mixed with water and sugar. There were of course occasional symptoms presented by the subjects in the studies of the Referee Board, not in the subjects of Professor Chittenden's studies who were fed baking powder food, but in those of Professors Long and Taylor, who were fed what were thought to be the baking powder residues; though with these subjects it was not till large quantities were administered that minor complaints appeared, attributed chiefly if not wholly to the action of the sodium sulphate. The Board very properly interpreted these findings as not indicative of any deleterious action. To have concluded otherwise would be to condemn any substance that might be unwholesome in inordinate amounts, that is, by what the writer has previously[4] termed an abuse of the quantitative relation. More than a hundred tons of s.a.s. baking powders have been consumed daily over a period of decades. It would seem that if there had been any actual digestive disturbance caused by a food so widely ingested over such an extended period of time, it would have been recognized and declared long before this and claims of its existence would not today rest solely upon speculation and questionable interpretations of experiments with substances other than the food itself.

XII

EVIDENCE AS TO THE ABSORPTION OF ALUMINUM

EXPERIMENTS ON HIGHER ANIMALS

In 1880 Mott announced (p. 47) that the aluminum of baking powders, instead of passing out of the system, accumulated in the various organs, interfering with their proper functions. This conclusion was based upon the result of the analysis of various organs of a dog that had been fed 16 oz. of aluminum hydroxide. His methods of analysis were not described, nor were any indications given of the exercise of proper care to exclude the introduction of aluminum through the reagents and dishes; nor were the actual amounts found described except for a general statement that aluminum was found in large quantities.

This was the state of the literature and this constituted the scientific "knowledge of the subject" when some twenty years later the writer undertook his investigations (p. 59) along somewhat similar yet different lines. Owing to the fact that the domestic hog is omnivorous and that his digestive organs and processes closely resemble those of man, it was regarded as a particularly suitable animal for the subject of such investigations. Further, since the real subject of interest was the aluminum residue of food prepared with s.a.s. baking powders, it was regarded as the only suitable procedure to feed the animal the actual food regarding which the inquiry was concerned. Again, it was believed that great emphasis was given to the baking powder residue by using 30 grams of a straight s.a.s. baking powder per 1000 cc. of cereal, flour and hominy in equal amounts. The feeding period, namely eighty-three days, during which the animal gained in weight from 73 lb. to 258 lb., was regarded as a sufficient period of time to constitute a rigorous test of the extent to which the body actually did take up and store aluminum. Considering the nature of the problem presented, it was thought that 50 gm. of tissue, less where necessary, was a sufficient quantity to yield the information sought.

The method of analysis employed (p. 73), was an application quantitatively of the accepted qualitative method of testing for aluminum, which was familiar to and used by the writer at that time. It is open to the objection that it employs sodium hydroxide for the separation of iron and introduces the necessity of avoiding glass dishes for the sodium hydrox- ide solution, and so is less convenient than methods developed in the intervening years. However, a supply of Kahlbaum's sodium hydroxide from sodium, free from detectable amounts of aluminum, was obtained and used in the analysis, as were also suitable dishes for the operations. The writer, as a student under O. D. Allen, one of the great analysts of his day, had taken a particular interest and had had special training in the detection of aluminum. This arose from the fact that the writer found aluminum almost constantly present in test material submitted to him by Professor Allen, so that under the latter's direction he had been led to follow out the method with unusual attention to details whereby the introduction of aluminum from reagents and glassware would be avoided. Yet, notwithstanding these facts, criticism of the method used by the writer has been made,[1] to the effect that caustic soda would necessarily dissolve aluminum from the glassware.

The witness when recently asked as to the use of caustic soda, some twenty years ago: "If he did not use glass vessels at all in connection with caustic soda would not that remove that element which you found to be most doubtful?" stated: "I cannot conceive of a chemist at that time using any other than glass utensils; they were not on the market; there was nothing to be used to handle it otherwise.

"Q. But, if he did not use glass vessels, does not that remove your criticism to a large degree?

"A. No, it does not. Even platinum is attacked, to a certain extent, by caustic soda; to a very slight extent.

"Q. Supposing he used platinum vessels, would that produce aluminum in his residue?

"A. It might.

"Q. If so-called silica dishes were used in the ignition of the organic material analyzed, might that not introduce a variable amount of aluminum in the final results that was not present in the original analysis?

"A. No, that would not have anything to do with it; it is entirely irrelevant; the question is absolutely meaningless, because it has no connection with what was being done there."

The following statement by the writer was written in the report of the work and published in 1907[2]: "During these chemical analyses great care was used to prevent the introduction of aluminum, both by the use of pure chemicals, previously tested as to the presence of aluminum and by the use of platinum and nickel dishes. The use of glass for alkaline solutions was avoided." As a matter of fact, the tendency of caustic alkalies to dissolve aluminum from glassware and the necessity of using suitable containers to avoid solution from glassware had been known for at least forty years, and for how much longer it has not seemed necessary to determine. The matter was not of sufficient importance to the case at issue to justify introducing evidence in order to correct it.

It is an interesting fact that other chemists have found Kahlbaum's sodium hydroxide free from detectable amounts of aluminum (p. 219).

As a result of the analysis of organs of the hog, as described, no aluminum was found. The writer concluded from these results that no aluminum was present. The results of the writer certainly indicated that aluminum was not present, as described by Mott from aluminum hydroxide ingestion by a dog. By the use of larger samples for analysis, possibly by a little better separation of the iron (though of this the writer is not fully convinced) and by the use of more delicate tests of the slight residue obtained (in most instances 0.1 or 0.2 mg.) there is a possibility of detecting by more recent methods amounts of aluminum less than would be indicated by the procedure used by the writer. Such smaller amounts would be less than 0.001 per cent in any event and in most instances less than 0.0001 per cent.

METHODS OF ANALYSIS

In recent analyses of blood and tissues for aluminum, one of two methods has generally been used:

1. Method adopted by the Association of Official Agricultural Chemists, modified by Steel.[3] The technique of securing the weight of the mixed aluminum and iron phosphates, is that of the A.O.A.C. Steel determines the iron content by the Mixer and DuBois[4] modification of the Zimmerman-Rheinhardt permanganate method in an aliquot portion of the original acid solution. A "blank" is made on the reagents used in the method and the "weight of the blank" is added to the amount of $FePO_4$; this weight is then subtracted from the combined weights of $AlPO_4$ and $FePO_4$. Following is the procedure as described by Kahn[5]:

Preliminary oxidation of organic material: "The weighed tissue was placed in a Kjeldahl flask and concentrated nitric acid solution added to it. The flask was heated slowly at first and then more vigorously until the solution became clear, when a moderate excess of nitric acid solution was added and the liquid boiled down to a small volume. The fluid was then treated with a fairly large volume of concentrated sulphuric acid solution for the expulsion of the nitric acid and for the complete oxidation of any residual organic matter. The sulphuric acid mixture was boiled for at least two hours after it became colorless, in order to eliminate NO_2. The residue was then dissolved in water, made up to volume, and the iron and aluminum determined."

Determination of Iron and Aluminum: "Obtain an aliquot portion of the available acid solution and remove any contained silica. Mix the liquid with sodium phosphate solution in excess of what is required to form normal aluminum phosphate. Add sufficient ammonium hydroxide solution to effect complete precipitation of the aluminum phosphate after thorough stirring. Then add hydrochloric acid solution, drop by drop, until the precipitate completely dissolves. Heat the liquid to about 50°C. and mix with it, at that temperature, a considerable excess of 50 per cent ammonium acetate solution and also 4 cc. of 80 per cent acetic acid solution. As soon as the precipitate of aluminum phosphate (mixed with iron phosphate) has sedimented, collect it on an ashless filter, wash it with hot water, ignite it and then weigh the residue.

"In an aliquot portion of the original liquid determine the amount of iron by the Zimmerman-Rheinhardt method. The calculated amount of $FePO_4$ is then subtracted from the weight of the mixed $AlPO_4$ and $FePO_4$."

2. Method of Schmidt and Hoagland.[6] The description is for the determination of aluminum in feces:

"Five to 10 gm. of feces are treated with several cubic centimeters of concentrated sulphuric acid and ashed in a silica dish. The soluble aluminum is dissolved out by warming with dilute hydrochloric acid. The residue on the filter paper is washed and then ignited and fused with sodium carbonate in a platinum crucible. The melt is dissolved out with dilute hydrochloric acid, the silica dehydrated, and the whole added to the main portion containing the aluminum. The volume at this point should be about 300 cc. and contain about 2.5 cc. of concentrated hydrochloric acid. Di-ammonium hydrogen phosphate is added to the solution, 0.5 gm. for each 100 mg. of aluminum phosphate present. The solution is heated, and while hot 5 gm. of ammonium thiosulphate (in solution) and after several minutes 6 to 8 gm. of ammonium acetate (in solution) and 4 cc. strong acetic acid are added. Heating is continued for about half an hour to expel SO_2, the precipitate is allowed to settle, filtered and washed once by decantation. The precipitate is redissolved in 2 to 2.5 cc. of concentrated hydrochloric acid, the solution diluted to about 300 cc., 0.5 gm. of ammonium phosphate added for each 100 mg. of aluminum phosphate present and the aluminum again precipitated as described above. The precipitate is filtered and washed several times with hot water to remove chlorides and ignited in a transparent silica crucible until constant weight is reached to remove excess of P_2O_5."

Several investigators have made comparative studies of the two methods. Paul E. Howe[7] concluded that the method of Schmidt and Hoagland commended itself from the point of view of manipulation. In the analyses of blood, the A.O.A.C.-Steel method yielded results slightly lower. L. J. Curtman and P. Gross[8] found the Schmidt-Hoagland method accurate both in pure solution and blood and advantageous from the point of view of technique, and the A.O.-A.C.-Steel method unreliable owing to the instability of $FePO_4$. Matthew Steel[9] found his adaptation of the A.O.A.C. method accurate when care was taken in the technique, but that the Schmidt-Hoagland method was as accurate and had the advantage of involving fewer manipulations.

C. A. Smith and Philip Hawk[10] found the Schmidt and Hoagland method more satisfactory, though the determinations in blood were not as accurate as was to be desired (8.75 per cent and 12.5 per cent high in duplicate determinations of known amounts added to beef blood. With correction by deducting the amount found in a "check" analysis of the beef blood, the excess became 5 per cent and 8.75 per cent).

A. K. Balls[11] gives the following estimate of the accuracy of the Schmidt-Hoagland method:

"The Schmidt-Hoagland method for the determination of aluminum gave results which involved a loss of as much as 7 per cent of the available aluminum, but which was usually about 4 per cent.

"The losses appear to have been due, in the main, to the formation of Al_2O_3 from $AlPO_4$, in the precipitate of the latter during ignition, but also partly to the solubility of $AlPO_4$ in the reagents and washings.

"The material, as finally weighed, is not wholly normal orthophosphate of aluminum, but contains less phosphoric anhydrid than does the same weight of orthophosphate.

"The indicated error might invalidate the method for accurate determinations of relatively large amounts of aluminum. For comparatively small quantities, however, the error appears to be negligible."

Balls advocates washing the precipitates with dilute ammonium phosphate solution instead of hot water to avoid loss and a final washing with ammonium nitrate solution to avoid excess of ammonium phosphate in the precipitate which requires very high and prolonged ignition to remove.

In 1915, Atack[12] published a method for the determination of small quantities of aluminum based upon the coloring of aluminum salts with alizarine s. Wolff, Voorstram and Schoenmaker[13] tried the method without success, probably because of their preparation of the dye, and following the suggestion of Keilholz,[14] substituted alizarine-sodium, made by dissolving pure sublimed alizarine (Kahlbaum) in just the right amount of Al-free NaOH and filtering several times. They destroyed organic matter in a quartz retort with sulphuric and nitric acids distilled from quartz retorts and not allowed contact with glass but kept in paraffined, or better, quartz bottles. In an examination of hen's egg, they found no aluminum in the white but from 0.02 to 0.06 mg. in the yolk. The following is the analytical procedure, as adopted and described by Philip P. Gray[15] of the Pease Laboratories Inc., New York:

"The digested (in some cases ignited) sample in the form of a sulphuric acid solution is evaporated in a platinum dish to small volume and then treated two times with ½ gm. potassium fluoride, evaporating each time to dryness to remove as much sulphuric acid and ammonium salts as possible. The residue is then taken up in ½ cc. concentrated HCl and 20 cc. H_2O. This solution is then transferred to volumetric flask and made up to volume, 50 or 100 cc. as the case may be, and then immediately transferred to a paraffined bottle.

"An aliquot of this solution representing a suitable charge of the original material is transferred to the platinum dish. About 3 cc. of 20 per cent sodium hydroxide (Kahlbaum from sodium, aluminum free) is then added, enough to make the solution strongly alkaline. The alkaline solution is diluted to about 15 to 20 cc. and heated to just below boiling. The solution is allowed to stand a few minutes and is then decanted and filtered into a Nessler tube in which has been placed a sufficient quantity of hydrochloric acid to neutralize the free alkalinity. The precipitate is washed several times in hot water. The filtrate, about 30 cc. in volume, is allowed to cool and 10 cc. of glycerine added. Then 1 cc. of a 1 per cent sodium alizarine solution is added; 5 per cent ammonia is next added, drop by drop, until the solution appears neutral as judged by the color of the indicator, and 1 cc. is then added in excess. After standing ten minutes, an amount of 5 per cent acetic acid is added, sufficient to neutralize the ammonia and yield a slight excess (the same quantity used in each case). The solution is then made up to 50 cc. The color is compared with aluminum standards made by using all of the above reagents; that is, sodium hydroxide, hydrochloric acid and glycerine as well as the alizarine, ammonia, and acetic acid, but containing in addition definite amounts of a standard solution of aluminum sulphate. Following is given the range of standards used, in which we found it best to work:

$$0 - .002 - .004 - .006, \text{ etc., up to .03 mg.''}$$

Subsequent to the conclusion of the writer that there was no accumulation of aluminum in the tissues of the body by absorption from the alimentary tract following the ingestion of food prepared with an s.a.s. baking powder (p. 87) and the demonstration of the fact that in 4 individuals the aluminum of such food was eliminated quantitatively in the feces (p. 90), the claims of Mott relative thereto have generally been disregarded. However, some ten years later, further investigations of the subject of the absorption of aluminum following the ingestion of such food were conducted in the laboratories of the Department of Medicine, Columbia University, at the suggestion of Professor Mallet, who for some years had appeared as an opponent of s.a.s. powders. The following are résumés of papers published about that time. In both investigations aluminum was determined by the A.O.A.C.-Steel method.

"ON THE ABSORPTION OF ALUMINUM FROM ALUMINIZED FOOD" BY MATTHEW STEEL[9]

Prof. J. W. Mallet of the University of Virginia suggested that Prof. Gies conduct experiments which would show definitely whether aluminum in aluminized food passes from the alimentary tract into the blood. Accordingly this investigation was undertaken at Prof. Gies' suggestion and under his personal supervision. The investigation consisted of 15 experiments on dogs, 11 pertaining to the passage of aluminum into the blood from alum and aluminized food, and 4 relating directly to the passage of aluminum from the blood into the feces. In experiments 1 to 7, in which alum was administered as a single dose in powdered form in "meat-hash pills," the following results were obtained:

TABLE LXXX

Al_2O_3 ingested as alum	Time hours	Al_2O_3 from blood analysis mg.
100	3	1.3
200	5	1.9
200	6½	2.6
300	5¾	6.7
400	7	5.9
400	7	10.9
400	6½	3

When similarly fed alum equivalent to 100 mg. Al_2O_3 daily for eight days, 1058 gm. of the blood of Dog 8, six hours after the last dosage, yielded 3 mg., as Al_2O_3. Dog 9, five hours after the last dosage yielded in 996 gm. of blood 4.3 mg., as Al_2O_3.

When fed daily for ten days an amount of s.a.s. baking powder biscuits, containing an amount of aluminum residue equivalent to 400 mg. Al_2O_3, Dog 10, weighing 13 kilos yielded in 650 gm. of blood 2.6 mg., as Al_2O_3; Dog 11, weighing 11.1 kilos, fed ⅘ the amount for nine days, yielded in 572 gm. of blood 1.5 mg., as Al_2O_3.

In any of the above experiments, prior tests for the presence of aluminum in the blood were not made, it being assumed that aluminum was not present.

"General Conclusions. 1. When alum was administered in aluminum-free food to dogs, or when dogs ingested biscuit baked with alum baking powder, aluminum in comparatively large amounts promptly passed into the blood.

"2. Absorbed aluminum circulated freely, but as it did not show any pronounced tendency to accumulate in the blood, its full effects must have been registered outside of the circulation.

"3. When aluminum chloride was administered intravenously, from 5.55 per cent to 11.11 per cent of the aluminum passed from the blood into the feces during the three days immediately after the injection. Whether the aluminum passed directly through the walls of the intestine or was excreted by the liver, or whether both channels (or others) were followed, has not yet been ascertained."

In the 9 experiments in which alum was fed the difference between total iron and aluminum and total iron, which difference was assigned to aluminum, amounted to an average of 1.02 per cent of the total combined weights; in the 2 experiments in which baking powder food was fed, it amounted to an average of 0.76 per cent of the combined weights. No qualitative tests were made to show that aluminum was actually present, and, of course, duplicate determinations were not possible.

"ON THE ABSORPTION AND DISTRIBUTION OF ALUMINUM FROM ALUMINIZED FOOD"[5]

Max Kahn, while a graduate student in the laboratory of Dr. Gies, fed 4 dogs biscuits made with s.a.s. baking powder (Bob White). All the animals seemed to thrive and gained in weight. At the end of the feeding, the anesthetized animals were bled to death and the organs analyzed for aluminum.

TABLE LXXXI

Animal	1	2	3
Duration days	61	55	52
Mg. Al_2O_3 of Baking Powder fed, daily average	20.44	14.32	23.95
Kg., Body weight, beginning	8.66	5.68	9.34
Kg., Body weight, end	10.72	6.64	10.5
Mg. Al_2O_3 per 100 gm. organs (parts per 100,000)			
Bile	13.0	2.7	20.9
Blood	1.7	1.1	1.7
Bone (femur)	1.6	...	2.7
Bone (skull)	0.0	0.0	0.0
Brain	0.0	0.0	0.0
Heart	0.0	0.0	0.1
Kidneys	6.0	1.2	2.5
Liver	3.6	5.5	1.8
Muscle	3.2	2.8	1.7
Pancreas	5.0	20.5	4.3
Spleen	8.0	6.3	11.1

In the urine of the fourth dog, female, weighing 13.85 kg. at the beginning and gaining thereafter, fed for nearly three months biscuit with daily average Al_2O_3–36.18 mgs., there was found 1.8 mg. Al_2O_3 per 100 cc. of urine. (Note. The published article describing this work states that the amount of blood of animal 1 actually used for a determination of the Al_2O_3 was 21 gm., this result corresponding to 0.36 mg. Al_2O_3.)

"*General Conclusions.* 1. When biscuits baked with alum baking powder are fed in a mixed diet to dogs, aluminum passes in considerable amounts into the blood.

"2. Such absorbed aluminum circulates freely and, although it does not show a tendency to increase proportionately in the blood, it accumulates to some extent in various parts of the body. The bile contains a particularly large amount of aluminum under such circumstances. The pancreas, spleen, liver, muscle and kidneys contain considerable amounts, while the brain and heart seem to resist accumulation of aluminum. The long bones, under the conditions of these experiments, contained aluminum. The flat bone of the skull did not contain aluminum.

"3. Aluminum, when ingested in aluminized food under the conditions of these experiments, is absorbed in part and is excreted, to some extent, in both the bile and urine."

Further investigations on dogs were made somewhat later. A. K. Balls, Graduate Student, working under Dr. Gies in the laboratory of Physiological Chemistry, Columbia University, conducted, as his major thesis for the degree of Doctor of Philosophy, 1917, certain investigations[16] on dogs both without and following the feeding of aluminum.

"No observations of the occurrence of aluminum in normal animal tissues have been recorded. It is probable that while any considerable quantities, if present, would have been discovered, at the same time traces might have been as easily overlooked, especially when the attention of investigators was not particularly directed to this end.

"Before determining whether aluminum is absorbed by the dog from food containing it in potentially soluble form, it was necessary therefore, to ascertain whether aluminum exists normally in dog tissues, and if so, to what extent.

"Accordingly, dogs were selected which had been fed for a period of about two months on a normal diet. This consisted of the following, grams per kilo body weight: hashed lean beef, 15; cracker meal, 2; lard, 3; bone ash, 1, samples of which constituents, when extracted with artificial gastric juice failed to give a filtrate containing aluminum.

"From 200 to 500 grams of blood were removed from each dog through a femoral artery, under local anaesthesia and the dogs afterwards maintained on the same diet. Three weeks later two of the dogs were bled to death, the operation being conducted as before. After exsanguination was nearly complete, physiological salt solution was admitted to a femoral vein, and the perfusion continued until practically all of the blood had been washed out of the body. The washings were discarded.

"The tissues were freed from organic matter in one of the following ways: 1. Decomposition of the material according to Neumann, with nitric and sulphuric acids; removal of any nitric acid by boiling with water, and evaporation of the excess of sulphuric acid from a silica or platinum dish. 2. Ignition in platinum, and subsequent fusion of the ash with sodium and potassium carbonates."

The residues were then analyzed by the Schmidt-Hoagland method. "In the case of bone, however, the fusion with alkali carbonate is not necessary. Here also the procedure of Schmidt and Hoagland must be modified. The large quantity of calcium phosphate present makes necessary a greater dilution of the solution for the first precipitation. Comparatively small samples of the material must also be used, even at the sacrifice of accuracy. Working with 2.5 gm. of bone ash, representing about four times that weight of fresh bone, the solution in which aluminum phosphate is first precipitated should measure about 800 cc. and the quantities of reagents be regulated accordingly. Otherwise calcium phosphate will contaminate any precipitate of aluminum phosphate. The following results were obtained:

"Data showing the aluminum content of normal dog tissues; Column I designates the animal used; Column II the part of the animal used; Column III the weight of the part in grams. Column IV the method of decomposition of the tissues; Column v the weight of $AlPO_4$ found* in milligrams; Column vi the aluminum calculated to Al_2O_3, and Column vii the aluminum calculated to milligrams of Al_2O_3 per 100 gm. of the original material.

TABLE LXXXII

I	II	III	IV	v $AlPO_4$	vi Al_2O_3	vii Mg. Al_2O_3 per 100 gm.
C.	Blood...............	588	g. Neumann	0.1 mg.	0.04 mg.	0.01 mg.
D.		513	Neumann	1.1	0.46	0.09
G.		330	Neumann	0.0		
H.		340	Neumann	0.0		
E.	1st bleeding.......	235	Neumann	0.1		
E.	2nd bleeding.......	445	Neumann	0.0		
E.	Spleen..............	13	Neumann	0.1		
E.	Kidneys (2).........	38	....	....	Determination lost	
E.	Bile + gallbladder ...	15	Neumann	0.0		
E.	Muscle (Leg)........	60	Neumann	0.0		
E.	Claws...............	3.5	Ignition	0.1		
E.	Bone, (Femur).......	10	Ignition	0.0		
E.	Liver...............	....	....	....	Determination lost	
F.	Blood 1st bleeding....	230	Neumann	0.1		
F.	2nd bleeding	320	Neumann	0.1		
F.	Spleen..............	10	Neumann	0.0		
F.	Kidneys (2).........	35	Neumann	0.1		
F.	Bile + gallbladder ...	6	Neumann	0.0		
F.	Muscle (Leg)........	60	Neumann	0.0		
F.	Claws...............	3	Ignition	0.1		
F.	Bone (femur)........	10	Ignition	0.1		
F.	Ferratin, from liver...	0.130	Ignition	0.0		
F.	Remainder of liver...	200	Neumann	0.0		

"In but one instance was the amount of aluminum phosphate large enough to exceed a reasonable error in making the weighings. In this case the blood undoubtedly contained aluminum, the presence of which was confirmed by the qualitative tests described

* Six blank determinations, using the same reagents and glassware were run parallel to those recorded here. They gave the following results in milligrams of $AlPO_4$:

Neumann Method	Ignition Method
0.2 mg.	0.3 mg.
0.3	0.3
0.2	0.3

The proper corrections have accordingly been subtracted from each weight recorded in this column.

later. Unfortunately, the animal died after the first bleeding and the body was inadvertently discarded before aluminum was known to be present in the blood. It was therefore impossible to continue work on the other tissues of this particular dog.

"We regard a contamination of the material during the course of analysis, and accidental errors in the manipulation, as very unlikely, but, of course, as possibilities. The suggestion is also to be considered that the animal, whose previous history is unknown, may have eaten aluminized food prior to the beginning of our experiment. The fact that this single result is directly at variance with all the others, leads us to regard it as exceptional, to say the least.

"It is reasonable to suppose that most of the elements could be found in living matter if our means for their detection were delicate enough. From these results, however, we may conclude that aluminum is present in the normal dog in amounts that are too small usually for detection."

The following experiments were then conducted, first, using a dog whose blood content of aluminum was previously found to be too small to be detected by the method used, fed for one month 10 gm. daily of moist aluminum phosphate in the place of bone ash of the normal diet; second, a similar experiment, using a young dog. Aluminum was found in the tissues as follows:

TABLE LXXXIII

| | Mg. Al₂O₃ per 100 gm. | |
	I	II
Blood plasma	0.09	0.00
Blood corpuscles	0.42	0.27
Bile and gallbladder	1.8	7.56
Spleen	0.0	0.97
Kidneys	1.58	1.20
Muscle (leg)	0.15	....
Bone (femur)	2.5	2.94
Bone marrow (femur)	5.8	4.20
Claws	0.0	....
Ferratin of liver	60.0	0.57
Remainder of liver	0.07 Liver	0.30

Ferratin from the liver of a "normal" dog contained 2.4 per cent iron (Fe_2O_3); from dog 1, 4.3 per cent iron (Fe_2O_3).

It was concluded that when fed to a dog as aluminum phosphate, considerable aluminum was absorbed; some is retained in the body, in all probability replacing the iron of ferruginous proteins (ferratin of liver) and accompanying the iron in its cycle (circulation in the body) bone marrow, corpuscles, liver, proteins, bone marrow. Still another portion remains in the bones. "There is," however, "a storage of aluminum, at least temporary, which is not confined to any one particular tissue. The distribution is not uniform, however, and seems to be roughly parallel to that of iron." "Much of the aluminum fed was speedily eliminated in the bile and in the urine."

Two dogs were fed for twelve weeks on a diet including baking powder biscuit made with Parrot and Monkey baking powder, so that each dog received daily about 20 mg. aluminum, as Al_2O_3 per kg. body weight and at the end was killed and the tissues washed by perfusion with physiological salt solution. The following amounts of aluminum were found.

TABLE LXXXIV

Animal	Mg. Al_2O_3 per 100 gm.	
	A	B
Weight, kg. at beginning...................	13.9	9.9
Weight, kg. at end........................	14.1	8.8
Blood....................................	0.06	0.30
Urine (in bladder)........................	0.35	1.65
Bile.....................................	0.93	12.55
Liver....................................	0.10	0.14

Balls reported that the dogs before they were killed were apparently all right. B was not a very vigorous animal and lost some weight, but there was not anything very remarkable about him.

John Allen Killian, associate professor of biological chemistry, New York Post-Graduate Medical School and Hospital,[15] fed biscuit made with Good Luck baking powder (straight s.a.s.) to a dog from October 13 to November 24, on which date the animal was killed and the organs analyzed using the Schmidt-Hoagland method, with the following results.*

TABLE LXXXV

Weight of dog, kg., October 7.............		......	....	8.6
Weight of dog, kg., November 24.........		......	......	7.9
Daily ingestion of aluminum from baking powder food, mg.				
as Al₂O₃..................		......	......	357
as Al....................		......	......	190
Aluminum in organs, as mg. Al₂O₃ per 100 gm.				
Blood, October 7,	Serum...........		......	0.08
	Clot..............		......	0.12
November 24,	Serum...........		......	0.25
	Clot...............		(.90)	0.10
Gallbladder and bile..................		(18.9)	(10.6)	1.67
Liver.................................		......	(1.05)	0.16
Kidneys..................................		......	......	0.44
Bones.................................		......	......	0.06

The dog was in good health during the feeding and up to the time it was killed. Shriveling of the stomach or intestines was not found, the mucosa (lining membrane) of these organs appearing normal, neither was there any corrosion, apparent to the naked eye.

Killian explained that although he had calculated the precipitates obtained as aluminum phosphate, against his best judgment, he did not regard the results as dependable or indicative of the absorption of detectable amounts of aluminum owing to the very small amounts of the residues obtained. He based this conclusion upon his experience with the method where he had used known small amounts of aluminum phosphate (1 mg.) and further upon the fact that resolution and reprecipitation of the residues obtained as aluminum phosphate yielded very much lower figures each time (in some cases more than 50 per cent) than the original

* In order that the results may be comparable with those of the previous investigators, the results have been corrected for the blank (0.2 mg.) and calculated to milligrams Al₂O₃ per 100 gm. material. In making this calculation in the case of the blood, recorded by volume (cc.), it has been assumed that the specific gravity of the whole blood was 1.060 and of the serum 1.026. The figures in parentheses are the preceding precipitations, reprecipitations having been made for purification.

weights obtained, while the loss with known aluminum phosphate similarly treated was only slight (about 3 per cent). Still further, iron was found present even in the residue after 3 precipitations, showing it was not pure aluminum phosphate.

Philip P. Gray,[15] using the modified Atack colorimetric method (p. 219), determined the aluminum in 2 presumably normal dogs (without feeding of baking powder food or of aluminum compounds) and obtained the fo lowing results (double figures are for Dogs 1 and 2 respectively, single figures are on composite samples from Dogs 1 and 2. Results are parts Al per million): blood 1.2–1.4; kidney 6–5.4, liver 5–7, gallbladder 4.1, spleen 2.2, testicle 1.0, pancreas 1.0, abdominal fat 2.8, bone marrow 7.2, thyroid, adrenals and mesenteric lymph nodes 18.8–16.3. On a third dog, likewise not fed aluminum in any form but fed biscuits raised with sodium acid sulphate and sodium bicarbonate: blood before feeding 1.0; after feeding, arterial 2.2, portal 2.7, spleen 4.6, abdominal membranes 2.4, thyroid 21., adrenals 14., mesenteric lymph nodes 2.7. Since blanks on the reagents were deducted, these figures appear to represent the normal or at least the incidental occurrence of aluminum in the organs of these particular animals, as determined by this method.

EXPERIMENTS ON MAN

Alonzo E. Taylor reported the work of the Referee Board relative to the absorption of aluminum, as follows:

"Steel reported the presence of a few milligrams in the blood of the dog, following the administration of large doses of aluminum. We have made the test upon the human being. Four subjects were given, for a period of several days, ingestions of aluminum amounting to about 1 gm. of the metal per day. From a vein in the arm of each man 200 cc. of blood were drawn. In this amount (over 800 cc.) it was not possible to demonstrate the presence of aluminum." (The Schmidt-Hoagland method was probably used.) "All reagents and apparatus (quartz and platinum) were demonstrated free from aluminum.

"Experiments were conducted to determine whether an increase of phosphate elimination, when the diet was poor in phosphate, demonstrated abstraction of phosphate from the body fluids and tissues such as might occur if aluminum were resorbed in a form other than phosphate, and eliminated as phosphate." (See p. 127.)

"The figures fail to substantiate the assumption that aluminum was resorbed and robbed the body of phosphorus and hence fail to give evidence of the resorption of aluminum. Resorption of traces we have obviously not disproved."

Experiments of Paul E. Howe and Louis Pine.[15] Louis Pine, in 1915, while pathological chemist to the City Hospital, at the request of Dr. Gies fed hospital patients, suffering with chronic diseases, biscuits made with a straight s.a.s. (30.75 per cent), baking powder (Parrot and Monkey brand), at the end of the feeding period collected the blood, between 10 and 11 A.M., which was subsequently analyzed for aluminum February to April, 1916, by Paul E. Howe, assistant professor of physiological chemistry, Columbia University, associated with Dr. Gies.

Prior to the feeding experiment the patients had been on an ordinary hospital diet, including ordinary bread; there were no special tests of the blood for aluminum. The collection of the blood was in Mason glass jars and the blood was stirred with a glass rod to prevent clotting. The drinking water was the New York City supply and was not tested for aluminum. The flour from which the biscuits were made and other parts of the diet were not tested for aluminum. For about one week they were fed 12 biscuits a day; as this seemed too much, it was reduced to 9 biscuits a day. The Schmidt-Hoagland method of analysis was employed. Sample D was lost. With each of the others a measurable amount of final precipitate was obtained which consisted largely of iron. None showed aluminum by the cobalt nitrate test; Sample B but not C showed a trace of aluminum by the Atack test, sensitive to one part in ten millions. The Goppelsroder test gave a positive reaction with Sample B. He concluded that a trace of aluminum was probably present in B, excluding contamination by the negative tests for aluminum in Samples A and C. There was no measurable aluminum in the reagents, as shown by blanks.

TABLE LXXXVI

Patient	A	B	C
Duration of feeding days................	62	54	30
Total aluminum from baking powder in biscuits fed, gm......................	24	21	11.6
Amount of blood, gm...................	643.7	505.7	283.4
Weights of final precipitates, mg.........	0.9	0.6	0.8
Amount of aluminum, in final precipitates	None	Trace (estimated mg. 0.0003)	None

Experiments of A. K. Balls and Samuel Gitlow.[15] Samuel
Gitlow, M.D., chief of the Department of Physiological
Chemistry, Lebanon Hospital, in 1916, at the request of
Dr. Gies fed convalescent patients at the Lebanon Hospital
bread made with a straight s.a.s. baking powder (Sunset
brand), collected the blood and urine and these were subsequently analyzed for aluminum by Arnold Kent Balls,
graduate student working under Dr. Gies in the laboratory
of Physiological Chemistry, Columbia University.

Prior to the feeding experiments the patients had been on the
average hospital diet; their blood was not tested for aluminum, nor
the city water drunk, nor the food they had been eating; nor did
witness know whether aluminum cooking utensils had been used
in the cooking of the food. The blood was withdrawn about two
hours after the meal.

The Schmidt-Hoagland method of analysis was employed.
The urine samples were divided into three equal parts and
aluminum determined on each of two of these portions. A blank
correction of 0.1 mg. was subtracted from each final weight of
aluminum phosphate obtained in all analyses.

The precipitates each gave positive qualitative analytical tests
for the presence of aluminum. The conclusions were that the blood
and urine undoubtedly contained aluminum.

There might be a little silica left in the final precipitate, but
Balls considered it unlikely in any considerable amount. From the
white color of the precipitates and their character, he would say
he had excluded the presence of iron.

TABLE LXXXVII

Patients	1	2	3	4	
				1	2*
Baking powder of bread, daily, approx. gm..................................	12	12	12	12	
Age, years.............................	38	45	..	27	
Duration of feeding, days..............	23	46	60	60	11
Amount of blood, cc...................	75	100+	lost	40	400
Balls' No...........................	#7	#6	..	#4	#5
Weight, gm..........................	....	116	..	58	475
Aluminum as Al₂O₃ mg. per 100 gm ..	0.05†	0.26	..	1.2	
Urine, period, days....................	....	3	..	3	3
Volume, liters.......................	....	3½	..	4	5
Balls' No............................	....	#3	..	#1	#2
Weight, gm..........................	....	2870	..	4090	3780
Aluminum, as Al₂O₃ mg. per 100 gm..	....	{ 0.01 / 0.02 }	..	0.12 / 0.11	

* Second feeding period, after an interval of four days.
† Contained in a bottle with an aluminum top.

Experiments of Philip B. Hawk and C. A. Smith.[15] These investigators withdrew blood from the 2 subjects who had been for six weeks on a mixed diet including food prepared with an s.a.s. baking powder (Seagull), at least twelve hours after the last of such baking powder food had been eaten, and examined both the blood so withdrawn, and the urine

TABLE LXXXVIII

Subjects	F	S
Amount of blood, gm.............................	666	795
⅖ employed for each determination..............	266.4	318
Mg. aluminum as Al₂O₃, obtained (Duplicates)....	{ 1.55 / 0.46 }	{ 0.38 / 0.84 }
Mg. per 100 gm. blood.........................	{ 0.58 / 0.17 }	{ 0.12 / 0.26 }
Rathgen and Atack qualitative tests..............	Positive	Positive
Urine (three days)		
Aluminum, as Al₂O₃, in ⅖ of total urine..........	{ 0.50 / 0.54 }	{ 0.17 / 0.17 }
(Duplicates) mg.............................		
Total per day, average, mg......................	0.433	0.14
Rathgen and Atack qualitative tests..............	Positive	Positive

collected in the fifth week, for aluminum, with the results Table LXXXVIII (weights reported as $AlPO_4$ are here calculated as Al_2O_3). (Bread eaten daily was 6 muffins or ½ loaf, containing 10 gm. of baking powder.)

The conclusion reached was that aluminum did not accumulate to any extent in the blood, but was present in small amounts as shown by the qualitative tests, indicating that it had been absorbed and that it was eliminated in the urine in small amounts.

The investigators concluded from their study of the method employed (Schmidt-Hoagland), that, as applied to blood, it was not as accurate as was to be desired and in some determinations gave variations as high as 1.5 mg., and that undoubtedly their determinations in the specimens of blood examined might be subject to the same ratio of error.

Prior to the experiment, the blood or urine of the 2 subjects was not tested for aluminum and so far as they knew aluminum might have been present. Possibly manganese, used in the process of analysis, might have been present in the residues weighed as aluminum.

Analyses of Willis G. Hilpert.[15] Clifton Evans, warden of the State Penitentiary of Arkansas, stated that twice daily during the past two years the occupants of that institution received food prepared with an s.a.s.-phosphate baking powder which they ate heartily and that the health of the prisoners was good; they generally improved in their physical condition while in the institution.

Paul L. Mahoney, physician at the State Penitentiary of Arkansas, testified that he took samples of blood from inmates of the institution, as described below.

Willis C. Hilpert, chemist of the Miner Laboratories, Chicago, testified that he received from Dr. Mahoney the samples of blood, as below, and that he analyzed them for aluminum by the Schmidt-Hoagland method.

The residues were examined qualitatively and found to contain sodium, potassium, calcium and aluminum. There also might be silica and titanium. The quantities of the residues were too minute to permit of any determination of the relative quantities of these various ingredients.

TABLE LXXXIX

Samples	A	B	C	D
Initials of subjects (2 individuals).....	C. F. J. J. A.	G. C. B. W.	E. B. C. B.	J. D. V. H. S
Weights of samples, gm..............	384.2	381.2	466.85	420.85
Weight (mg.) of residue, as aluminum phosphate........................	1.2	1.1	1.9	1.8
Iron (mg.) of residue...............	0.07	0.13	0.1	0.2
Weight (mg.) of blank..............	0.3	0.3	0.3	0.3
Corrected weight (mg.) of residue.....	0.83	0.67	1.5	1.3
Corrected residue (mg.) per 100 gm. blood...........................	0.21	0.17	0.32	0.30
Calculated as Al_2O_3................	0.087	0.07	0.13	0.12

In a commercial analysis, he would be quite likely to disregard altogether such quantities as were found. It was possible that the aluminum found qualitatively in the residues might have come from the reagents.

Analyses by Frank C. Gephart.[15] Mrs. Laura K. Brightman of Talladega, Alabama, manager of the Central Hotel, testified that for four months her housekeeper, Mrs. Watts, and for two years her pastry cook, Lizzie Tanner, and herself had eaten all their meals at the hotel and partaken of food at each meal prepared with baking powder, until recently Calumet brand (s.a.s. and phosphate) and since her last purchase Vision brand (s.a.s.).

Ernest K. Thornton, chemist of the R. B. Davis Baking Powder Company, testified that he obtained samples of blood from Mrs. Brightman, Mrs. Watts and Lizzie Tanner and delivered the same to Dr. Gephart.

Dr. P. H. Brigham, commissioner of health of Florence, South Carolina, obtained samples of blood in paraffined bottles from 3 attendants of the South Carolina State Hospital, in Columbia, who had eaten food at that institution prepared with baking powder containing s.a.s. for periods of five and a half years, four years and eight months, respectively. These were sealed and forwarded to Dr. Gephart.

Frank C. Gephart, chemist and formerly associated with Dr. Long in the conduct of chemical work for the Referee

Board, testified that he had received the above blood samples and analyzed them for aluminum using the Schmidt-Hoagland method. The samples from Dr. Brigham were numbered 1, 2 and 3; those from Mr. Thornton, 4, 5 and 6.

TABLE XC

Number of samples	1	2	3	4	5	6
Initials.............	H. L. W.	R. B. M.	G. A. F.	B	LT	W
Weight of blood, gm..	455	460	443	343	347	330
Residues, as AlPO$_4$, mg..............	12.5	8.4	11.4	13.3	13.2	6.7
Blank, as AlPO$_4$, mg.	6.7	6.7	6.7	6.7	6.7	6.7
Alizarian test for Al..	+	+	+	+	+	+
Residue, less blank, mg..............	5.8	1.7	4.7	6.6	6.5	0.0
As mg. Al$_2$O$_3$ per 100 gm. blood......	0.533	0.155	0.444	0.806	0.784	0.0
Alizarin test for aluminum.........	+	+	+	+	+	+
Iron...............	+	+	+	+	+	+

The residues were not tested for sodium, potassium, calcium, or magnesium. Such contaminations were possible and would have much greater influence on very small amounts than on larger amounts for which the method was designed. The aluminum of the blanks might have come from the reagents or utensils. The qualitative tests for aluminum were as strong on the blanks as on the blood residues, so that there did not appear to be more aluminum in the blood residues than in the blanks. "I do not think there was any aluminum in the blood. I could not say conscientiously that they did contain aluminum."

M. Gonnermann,[17] using a method that does not inspire confidence, has reported the wide occurrence of aluminum in foods, in calculi and in the organs of man. He attributes its presence in the body to absorption from both naturally occurring and added aluminum compounds in food.

Opinion of Andrew J. Patten.[15] Andrew J. Patten, chief chemist for the Michigan Agricultural Experiment Station and for five or six years Referee of Methods of Analysis for Inorganic Constituents of Plant Material for the Association of Official Agricultural Chemists, stated that one of the chief problems in connection with that work had been the determination of iron and aluminum in the presence of a large amount of phosphate, such as in the ash of seeds, and in fact of all agricultural material.

Every method he could find in the literature had been considered. In the material examined, besides aluminum, there had always been small amounts of other material as well as iron, calcium and magnesium. Almost any material which was used to precipitate small amounts of aluminum or iron would precipitate the other in varying degree. The gelatinous character of the aluminum precipitate rendered it difficult to prevent contamination with calcium phosphate. He had never found a method that would completely separate iron from aluminum. Repeated precipitation would reduce the amount of iron, but errors were continually being introduced. This was true of the thio-sulphate (Schmidt-Hoagland) method, both in his experience and that of the U. S. Bureau of Standards.

It was his opinion that it would be almost impossible to determine such a small amount as 1 mg. of aluminum phosphate in any material comprising 400 gm. and do it accurately. It would be impossible to establish absolutely that that amount did not come from contamination. In their work they ordinarily considered 1 mg. not worth their attention.

XIII

THE EFFECTS OF ALUMINUM COMPOUNDS WHEN ADMINISTERED SUBCUTANEOUSLY OR INTRAVENOUSLY

EARLY STUDIES

The early studies of the action of alum, directly introduced into the tissues or circulation of the body, always in massive doses, do not call for serious consideration because of their very evident inapplicability to the subject at hand. Thus, Orfila's observation[1] in 1848 that 32 gm. of alum applied beneath the skin of a dog inflamed the parts so that, as he says, "the animal may succumb at the end of fifteen or twenty days as a result of the abundant suppuration which terminates the inflammation" has only historical interest. An experiment of this nature, before the days of aseptic surgery, does not now indicate any toxic action of the alum whatsoever, but an operative infection such as may well have occurred entirely independent of the alum. On the other hand, it is not intended to deny the possibility or probability of harmful effects of such treatment. Whatever effect alum so applied might have is probably no more serious than would be produced by the corresponding salt of iron; in neither case does it prove anything relative to the ingestion of the substance as it occurs in food.

Likewise, Pereira's observation[2] in 1852 that "after its absorption, alum appears to act as an astringent or astringent-tonic on the system generally, and to produce more or less general astriction of the tissues and fibers, and a diminution of secretion" affords no real knowledge of the subject in the present day, when it is recognized that opinions in order to justify themselves must have the support of accurate knowledge.

237

These early studies are mentioned here because only recently they and similar matter have been offered in evidence and cited in argument in support of the claim that aluminum in food is injurious to man since it thus gains entrance into the body. Unfortunately, the uninformed reader may be misled by such citations, the very presentation of which is unfortunate in any serious attempt to establish the facts relative to the action of food.

LATER STUDIES

The studies of Siem published in 1886 (p. 47) are the first of any investigations of more than historic interest on the effect of aluminum compounds administered either subcutaneously or intravenously. However, as already noted, his demonstration of the lethal doses for various animals of the double tartrate of aluminum and the production of certain gross lesions, notably degenerative changes of the liver and kidneys, are seriously impaired by the fact since demonstrated that tartrates alone are fully capable of producing at least some of the lesions found, and leave it uncertain as to what extent the effects obtained are attributable to aluminum.

Döllkin's work published in 1897 (p. 56), the next succeeding investigation of any importance on this subject, unfortunately is open to the same objection of uncertainty as to what extent lesions are to be attributed to aluminum or tartrates, since he likewise administered the double tartrate of aluminum. His exact studies of the microscopic pathological histology of the central nervous system are a valuable contribution to our knowledge of the lesions produced by aluminum tartrate and offer an explanation of the symptoms attributable to the nervous system observed by both Siem and himself.

Action of Tartrates[2a]

At the time of the work by Siem and by Döllkin, the toxic effects of tartrates was unknown. In 1912, following the observation of Baer and Blum[3] that sodium tartrate subcutaneously injected may greatly diminish the output of nitrogen and dextrose in the urine of phlorizinized dogs, F. P. Underhill[4] found that such administration of sodium tartrate induced disintegrative changes in the kidney tubuli, sufficient to account for the lessened elimination. The histological work was done by H. Gideon Wells of Chicago. Later Underhill, Wells and Goldschmidt[5] demonstrated that in the rabbit, the disintegrative influence of sodium tartrate upon the convoluted tubules was sufficient to prevent the elimination of tartrate by the kidneys. The same investigators, reviewing and extending the former observations,[6] summarized their conclusions, as follows:

"Salts of tartaric acid administered subcutaneously to fasting phlorizinized animals exert a markedly detrimental influence upon the secretory efficiency of the kidney, which is indicated by a greatly lessened output of certain typical urinary constituents. A histological study of the nephritic kidney demonstrates that the salts act specifically upon the epithelium of the convoluted tubules, and to a less extent upon the tubules of the loops of Henle, the glomerules, and interstitial tissue remaining unharmed. In the disintegrative process taking place, vacuolation first occurs, is rapidly followed by necrosis, and finally the dead cells or their debris may entirely fill the lumina of the tubules and form granular and hyaline casts.

"There is no strict relation between the dose of tartrate and the extent of damage inflicted. While large doses invariably induce a well-marked response, small doses may at times produce effects equally significant.

"Tartrates introduced into fasting animals call forth symptoms practically identical with those observed in fasting phlorizinized animals. It is therefore apparent that in the establishment of the pathological condition under discussion phlorhizin is without significant influence.

"Neither the liver nor the adrenal exhibits any detrimental effect from the injection of tartrates.

"The introduction of tartrates by way of the mouth to fasting rabbits is not nearly so effective in the production of nephritic symptoms as the administration of much smaller doses subcutaneously. In general, under the former circumstances the initial stages only of epithelial disintegration of the convoluted tubules obtain, which, however, are scarcely sufficient to account for the rapidity with which death usually ensues.

"Although in well-fed animals distinct pathological changes in the kidney are induced by the introduction of tartrates *per os*, these abnormalities are less in degree, but similar in kind, than those provoked under like conditions in the fasting animal. When tartrates are given subcutaneously to well-fed rabbits the effects evoked are somewhat less pronounced than when the salts are injected into fasting animals. From these facts it is evident that the state of nutrition plays a part in the development of tartrate nephritis.

"It is indicated that the introduction of a sufficiency of alkali to animals in a state of fasting permits a greater elimination of urinary constituents during tartrate nephritis than obtains under similar circumstances when the alkali is omitted. Histologically there is evidence that the administration of alkali exerts a slight modifying action."

Recently, G. E. Simpson[7] has shown that the urinary elimination of tartrates ingested orally by the rabbit was considerably less than in the case of the dog, the evidence supporting the probability that the difference was to be ascribed to greater bacterial destruction of tartrate in the intestine of the rabbit. Hence, where there was less bacterial destruction in the intestine and accordingly more tartrate was absorbed, there would be relatively greater toxicity from oral ingestion.

ELIMINATION IN FECES

It is of some interest to note the following report of Matthew Steel[8]: In 4 experiments in which aluminum chloride in dilute solution in amount equivalent to 9, 18, 36 and 82.8 mg. Al_2O_3, respectively, was injected slowly into the saphenous veins of dogs that for about ten days had been on a standard aluminum-free diet, there was recovered in the feces, during the three days following the injection, aluminum in the amounts 1.0, 1.1, 1.9, 4.6 mg. as Al_2O_3, respectively, being 11.11, 6.11, 5.27, and 5.55 per cent of the respective amounts administered.

Subcutaneous Administration to Mice

Recently, Lester C. Himebaugh,[9] of the Pease Laboratories, New York, has reported results of testing the toxicity of aluminum by the phenol toxicity method of the United States Hygienic Laboratory. It will be recalled that the original purpose of this method was to compare with carbolic acid as the standard the action of substances of a similar general character. Its value for expressing any relative degree of toxicity of substances such as salts of aluminum in terms of carbolic toxicity has never been demonstrated and is open to grave doubts. In this procedure, solutions of aluminum salts were injected under the skin of white mice and the dose per gram of body weight of the animal that was fatal in twenty-four hours was determined and compared with the dose of carbolic acid that similarly proved fatal. The following results were obtained:

TABLE XCI

Substance	Lethal dose, gm. per gm. body weight	Phenol toxicity coefficient, per cent
Ammonium aluminum sulphate, cryst........	0.004	11.2
Potassium aluminum sulphate, cryst..........	0.008	5.6
Sodium aluminum sulphate, cryst............	0.008	5.6
Sodium aluminum sulphate, anhydrous.......	0.004	11.2
Aluminum sulphate........................	0.004	11.2
Aluminum chloride........................	0.008	5.6
Carbolic acid.............................	0.00045	100.0

Intravenous Administration to Rabbits

The effects on the circulation of the intravenous administration of aluminum compounds into rabbits has been the subject of investigations by Florence B. Seibert,[9] of the Otho S. A. Sprague Memorial Institute of Chicago, Pathological Department, of which Dr. H. Gideon Wells is director. The work was undertaken at the suggestion of Dr. Wells and the results were offered by the respondents in Federal Trade Commission, Docket 540, as part of the evidence that s.a.s. baking powders are injurious to health. If one disregards the materiality of the observations to the question at issue, the results are of value and interest as indicating the effects upon the rabbit of aluminum salts, as administered.

To each of 5 rabbits, 5 cc. of an 0.8 per cent solution, being 0.04 gm. of soda alum (crystallized sodium aluminum sulphate, $Na_2Al_2(SO_4)_4.24H_2O$) was administered daily by intravenous injection for periods varying from thirty to fifty days. At the site of injection, swelling, edema, inflammation and necrosis, even to the extent that a portion of the ear sloughed off, were produced. Some of the animals lost weight; all developed rough coats and exhibited tremors. The dosage was based upon Kahn's declaration that he found 1.7 mg. Al in the blood of a dog.

Blood Changes. The changes observed in the blood were:

1. Reduction in hemoglobin. The normal range of the blood hemoglobin of the rabbit was stated to be from 70 to 90 per cent (determination by the Sahli hemoglobinometer). At the outset of the treatment, the average hemoglobin for the 5 rabbits was 83 per cent. This diminished rapidly till at the end of thirty-nine days it was 46 per cent. Following the discontinuance of the injections there was some increase of the hemoglobin.

2. Decrease in the number of erythrocytes. The normal number of erythrocytes per c.mm. of blood was stated to be about 5 millions. At the outset of the treatment, the average number for the 5 rabbits was 5.4 millions. At the end of forty days it was about 3 millions. Following the discontinuance of the injection there was some increase in the number of erythrocytes.

3. Changes in the morphology of the erythrocytes. At the outset, the blood did not present abnormalities in morphology. After forty days of treatment there were present abnormalities of shape, irregularities in size and polychromatophilia. Erythroblasts were also found in the blood, as also an occasional erythrocyte exhibiting stippling.

4. Absolute increase in the number of leucocytes. The average normal variation in the total number of leucocytes in rabbits' blood was stated to be from about 8000 to 12,000 per c.mm. At the outset of the treatment, the average number found was 11,250. After twenty days of treatment the number of leucocytes had

increased to 13,000 which was about the number found throughout the remainder of the treatment.

5. Change in the relative number of the polymorphonuclear and the small mononuclear leucocytes. The percentage number of these two types was stated to total nearly 100. At the outset of the treatments the polymorphonuclear leucocytes numbered 65 per cent and the mononuclear leucocytes numbered 27.5 per cent. With the progress of the treatment there was a gradual diminution of the former and increase of the latter, till on the twenty-seventh day they were present in about equal numbers (47 per cent each). Thereafter, the number of mononuclear leucocytes exceeded the number of polynuclear leucocytes; at the end of fifty-four days the relation was 50 per cent of the former and 44.2 per cent of the latter. This change, which was stated to be the reverse of what would be expected from an inflammatory leucocytosis, was regarded as showing that the absolute leucocytosis was not due to an infection in the animal.

6. An increase in the resistance of the red blood corpuscles to hemolyzing agents. Two hemolyzing agents were used, water and saponin. The test with water was made by suspending the corpuscles in a 0.8 per cent sodium chloride solution and adding water until a dilution was reached at which the corpuscles were completely hemolyzed. At the outset, the addition of water to produce a dilution of the solution from 0.8 per cent salt content to 0.41 per cent completely hemolyzed the corpuscles. After eleven days, greater dilution was required to produce this result, the maximum dilution required, namely to a salt content of 0.3 per cent, being reached on the thirty-sixth day. On the fifty-sixth day the dilution required was to a salt content of 0.335 per cent.

In making the saponin test, the number of cc. of a 0.01 per cent solution of saponin required to hemolyze completely a fixed volume of suspended corpuscles was the measure of the resistance to hemolysis. At the outset this was 0.35 cc. There was little change until after the thirty-sixth day when the amount required was 0.41 cc. Determinations on the forty-fifth day required 0.55 cc.; and on the fifty-second day, 0.98 cc.

The results were interpreted as showing that the red blood corpuscles were more resistant, and tougher than they were before receiving the injections. In certain anemias this phenomenon was stated to accompany an increased fragility of the corpuscles; that is, although they seemed to be less easily hemolyzed they were more easily broken down by mechanical friction.

A second series of observations was made on rabbits, each of whom received daily by intravenous injection 0.01 gm. aluminum chloride ($AlCl_3.6H_2O$). It will be noted that whereas the amount of aluminum injected as soda alum amounted to 2.4 mg. daily, the amount injected as chloride was 1.1 mg. daily or slightly less than one-half the previous dosage. The effects obtained were of the same order as those obtained with the alum as described above, excepting to a less extent. A control series of observation was made on rabbits by daily intravenous injection of 5 cc. of sulphuric acid of pH 3.4, the hydrogen-ion concentration of the soda alum solution employed in the first series. It was reported that the results were negative; that is, that after forty days of such treatment the hemoglobin and erythocytes were not diminished, the leucocytes were not increased, the relative number of the mononuclear and polymorphonuclear leucocytes was not changed and there was no increase in resistance to hemolysis. Moreover, there were none of the changes in the morphology of the red corpuscles that were noted in the first and second series. The results seemed to indicate that the action of the aluminum salts was not due to the hydrogen ions present in the solutions of these salts that were injected.

A third series of experimental observations was made on 4 rabbits, to whom soda alum was administered daily, mixed with lactose in gelatin capsules, to the amount of 0.1 gm. The hemoglobin at the outset averaged 85 per cent; at the end of the period of thirty-nine days, 77.5 per cent. Other changes in the blood did not occur; that is, the erythrocytes remained normal in number and morphology; there was no leucocytosis produced; there was no inversion in the relative occurrence of the polymorphonuclear and the mononuclear leucocytes, but rather a relative increase of the former and diminution of the latter; and there was no change in the hypotonic salt or the saponin hemolysis. Notwithstanding that the decrease of hemoglobin was slight and within the range of normal variations, since the variations did not occur from day to day and the decrease was gradual, the hemoglobin reaching its low point in the last determination, the investigator thought she could say that the effect was due to the feeding and not to the normal variation of the animals or to confinement of the animals.

The decrease, however, was "merely suggestive." It was true that the decrease was not absolutely gradual, the curves of the individual animals at times showing an increase but never extending up to the original.

Effects on Anti-Body Production. Dr. Siebert has also reported[9] some preliminary tests for the determination as to whether the injection of aluminum solution had any effect upon the resistance to infections of the animals; that is to say, the effect upon anti-body production, specifically, hemolysin and agglutinin.

In a first group of observations the amount of hemolysin and agglutinin production from the injection of suspended sheep red corpuscles was determined. These results constituted the normals or controls with which the subsequent results were compared. It was found that where the intravenous injection of sheep red corpuscles was followed by the intravenous injection of the soda alum solution, the latter was without effect upon hemolysin and agglutinin production. When, however, several injections of the soda alum solution preceded the injections of the sheep red corpuscles suspensions, there was no production of hemolysin, and agglutinin production was less than in the controls. If, however, there were as many as 30 to 50 consecutive (presumably daily) intravenous injections of the soda alum solution, hemolysin production resulted from the injections of the suspended sheep corpuscles, but to a less extent than in the control animals; while the serum of animals so treated agglutinated sheep red corpuscles to a very marked extent. It was suggested that this latter result might be explained by the facts that there was a slight amount of aluminum present in the serum and that aluminum is known to agglutinate red blood corpuscles; and that the type of agglutination observed with the serum of these animals was different from that with the serum of the control animals.

From these preliminary experiments the investigator did not think that the production of protective sustances was increased by prior injection of soda alum; on the contrary, it looked as if both the hemolysin and agglutinin production was decreased, the apparent increase of agglutinating power being in reality an effect due to the presence of aluminum in the serum, rather than an increased production of agglutinin.

XIV

EXPERIMENTAL OBSERVATIONS UPON THE INFLU-
ENCE OF FOOD PREPARED WITH BAKING POWDER
CONTAINING S.A.S. UPON THE GROWTH AND
GENERAL WELL-BEING OF ANIMALS AND
OF MAN

Under the subject matter of this chapter should be included certain of the investigations of E. E. Smith, reported in Chapter v, and of the Referee Board, reported in Chapter vi.

Attention has already been called to the existence of the claim that alumed flour was injurious to health; and further, to the fact that this idea was not based upon any demonstrated injurious effects from the ingestion of such flour but on a general idea that because alum in large dosage could act injuriously, so in smaller amounts it would act similarly, only to a lesser degree. The addition of alum to flour has now been abandoned so that we are not directly concerned with the question of the influence of such food on growth and general well-being. Because the practice was reprehensible, being admittedly sophistication, no challenge of the claims of unwholesomeness was made. This is mentioned merely to indicate that opinions thus expressed cannot be given the weight of authority in criticism of present-day usages. The problem should and in the final analysis will be decided upon the merits, as determined by exact and impartial clinical observation and scientific demonstration. The only interests that should enter into its solution are the hygienic and economic welfare of the consuming public.

In dealing with the subject of this chapter, it is well to begin by again considering what is meant by unwholesomeness or injury to health. Does it mean that if a substance be given at any time and in any possible way and in any

amount and disturbances arise therefrom, then that substance is to be declared unwholesome? Or is the question
whether or not the substance does in fact act unfavorably to
the physical well-being of the consumers as it is used in food
in everyday practice? If the first viewpoint is to be the
criterion of judgment, then of necessity it must similarly
follow that any substance may be declared unwholesome
since, so far as the writer knows, there is no food or food
ingredient, however necessary or advantageous to the body
or however pure as to quality, that may not be abused at
some time or in some way or by ingestion in some quantity.
If such a viewpoint is not to be the criterion of judgment,
then we may adopt the second viewpoint, and the interpretations of disturbances observed or produced experimentally
will of necessity take into consideration the question as to
whether such disturbances have arisen or will arise in the
practical, daily consumption of the food in question.

In experimental work it is not only justifiable but important to study the effects of the subject of investigation
under unusual and extreme conditions, so as to direct attention to what the action might be if it were injurious. The
determination of such facts, however, does not prove that
the substance is injurious since it does not constitute the
solution of the problem of what occurs in the practical, daily
use of such food. Indeed, if it be discovered that effects by
abuse are not produced in actual consumption, then the
logical and only conclusion indicated is that the food is not
unwholesome or injurious and that like any other food it
may become so only because of its abuse. Unfortunately,
experimental results under conditions that must unquestionably be considered as an abuse of the food under investigation are sometimes interpreted as proof that under the
entirely different conditions of practical consumption the
food is likewise to be regarded as unwholesome or injurious.
Such an attitude is both erroneous in its interpretation of the
facts and unfair in its estimate of the particular food under
consideration. Its general application would lead to condemnation so comprehensive as to include practically all foods.
In many instances experimental results so obviously come

within the characterization of abuse in experimental administration as to be self-evident.

Following the announcement of the results of the investigations of the Referee Board, Leary and Sheib[1] studied the effect of the ingestion of aluminum upon the growth of the young. The object of this work was to ascertain the possibility of inhibiting the growth of young animals by diverting the phosphate content of the food to the intestine by means of an aluminum compound and to determine whether aluminum reabsorbed under these conditions. The experimental animals (puppies, young and full-grown white rats) were fed on a diet of low and known phosphate content; $Al(OH)_3$ was the aluminum compound used. In the case of the puppies the excreta were analyzed for phosphate at the end of the experiments, the animals were killed and the tissues examined with special reference to the presence of aluminum. The addition of aluminum to the diet of the puppies increased the fecal excretion of P_2O_5 and decreased the amount excreted in the urine in general in proportion to the amount of aluminum ingested. At the same time there was a decrease of titratable acidity of the urine. With amounts of aluminum up to 150 mg. daily (1000 mg. per kg. body weight or 1000 R.B. units) for about four months, the growth of young rats receiving a diet with a fixed amount of phosphate was not perceptibly influenced. With larger amounts, the body weights tended to decrease during the latter part of the experimental feeding. Absorption of aluminum occurred both in the dogs and the rats, the liver being the site of greatest deposition. This work confirms the statement of the Referee Board with reference to the deflection of phosphate, but of course has no special bearing on the question of the behavior of aluminum compounds as used in the diet of man, as the amounts of such compounds used were excessive (daily amount about 100 mg. to 300 mg. per kilogram body weight or 100 to 300 R.B. units in the case of the puppies and 500 to 3000 mg. daily per kilogram of body weight or 500 to 3000 R.B. units with the rats). In the diet of man, "even a very excessive use of aluminum compounds could scarcely furnish an amount of the metal in excess of 4 mg. daily per kilo of body weight" (4 R.B. units).

Recently the results of experiments by Hattie L. Heft, Maxwell Karshan and William J. Gies, under the title "Sanitary Studies of Baking Powders. Food made with alum baking powder is unwholesome" have been submitted in evidence.[2] Referring to the statement in the announcement of the results of the investigations of the Referee Board of Consulting Scientific Experts, Bulletin 103, United States Department of Agriculture, April 29, 1914: "In short, the board concludes that alum baking powders are no more harmful than any other baking powders," and commenting that "this conclusion appears to be based, in part, upon a failure to find that aluminum had been absorbed into the blood of healthy young men to whom aluminum in some form had been given in their food, or to note any other deleterious effect than catharsis from large doses of baking powder," these investigators prepared "to remove, if possible, the doubts created by the Referee Board's report on alum in foods, and give direct matter-of-fact answers to the questions:

"1. Whether alum baking powder is more harmful than any other baking powders, or

"2. Whether bread made with alum baking powder is as wholesome as that made with yeast, or with such baking powders as those containing tartrate or phosphate."

They proceeded on the basis of the three following general considerations:

"1. The relative dietetic values of different baking powders can be determined directly by comparing the nutritive quality of biscuits made with yeast with biscuits made in the same way with baking powders.

"2. The direct relative nutritive efficiency of yeast biscuits and 'baking powder' biscuits can be shown promptly by feeding as much as possible of each of such kinds of biscuits to growing animals, to the exclusion of as much as possible of all other kinds of food, without stopping the growth of the animals on the 'control' biscuits.

"3. Relative nutritive qualities of baking powder biscuits can be shown clearly by comparative effects of such food on the nourishment of pregnant animals and their offspring."

Young albino rats were used and were housed and cared for under conditions regarded as suitable.

Experiment I. Comparison of yeast with baking powder biscuits. Subjects, 6 male albino rats in each of 4 groups.

RECIPE FOR BISCUITS

1 cup flour (111.6 gm.)
⅓ tablespoonful sugar (4.9 gm.)
⅓ tablespoonful fat (4.2 gm.)
⅓ teaspoonful salt (1.8 gm.)
⅓ cup liquid: half milk (41.2 gm.), half water (35.5 cc.)
*2 *level* teaspoonfuls baking powder:
   Alum... 7.3478 gm.
   Phosphate (A).............................. 6.7024 gm.
   Tartrate (A)................................ 8.2208 gm.

* On the twentieth day, increased to 3 teaspoonfuls.

During the first thirty days of the biscuit period, the rats were given equal amounts of biscuit and milk. During the remainder of the biscuit period, the regular dietary plan was: biscuits alone on week days, cabbage and cheese on Sundays.

TABLE XCII

SUMMARY OF DATA, EXPERIMENT I

Figures are average weights (grams) per rat of the particular groups

Periods	Kinds of Biscuits			
	Yeast	Alum	Phosphate A	Tartrate A
End of fore................	41	46	41	41
Biscuit				
Thirteenth day..........	77	78	75	76
Thirty-second day........	121	117	118	115
Fifty-fifth day...........	147	122	136	138
Gain....................	106	76	95	97
After				
Seventy-second day.......	251	237	260	254
Gain...................	104	115	124	116
Average daily amount (grams) of food eaten				
Biscuit and milk...........	22.8	22.7	22.8	22.7
Biscuit....................	17.0	13.8	14.9	14.6
After.....................	34.3	32.8	34.1	33.8

The investigators ascribed the eating of less food by the animals of the alum group to "accumulative *toxic action* in the biscuit period, and to *delayed recovery* from that toxic action in the after period." They expressed the belief that there was "no reason to believe that a difference in palatability, against the alum biscuits, accounted for the effects observed."

All animals were normal in all respects throughout the experiment, excepting that those of the alum group showed:

Occasional diarrhea, usually mild, in a few instances pronounced.

On the thirty-third day, 5 of the 6 rats exhibited discoordinated movements, tremors, awkwardness in the use of the hind legs, with pronounced diarrhea. The symptoms gradually diminished and were absent five days later. Three and four days subsequently, tremor and paralytic conditions reappeared, followed by weakness and somnolence; gradual recovery during the following week. Two days later the symptoms appeared in relatively slight degree lasting only a few hours. No symptoms during the four subsequent days, which ended the period. The experimenters regarded the above symptoms as those of aluminum poisoning.

Experiment 11. Baking powder biscuits. Subjects, 6 female albino rats in each of 4 groups.

RECIPE FOR BISCUITS

> 1 cup flour (111.6 gm.)
> 1 tablespoonful fat (12.6 gm.)
> ⅓ teaspoonful salt (1.8 gm.)
> ⅓ cup liquid: half milk (41.2 gm.), half water (35.5 cc.)
> 3 level teaspoonfuls baking powder:
> Alum.................................... 11.0217 gm.
> Phosphate A............................. 11.0536 gm.
> Phosphate B............................. 12.3312 gm.
> Tartrate................................ 12.3312 gm.

The so-called alum powder used was Success brand.

During the first eighteen days of the biscuits period, the rats were fed exclusively on biscuits, without Sunday interruption with feeding of cheese and cabbage; subsequently, the Sunday feeding of cheese and cabbage was followed. On the second day of the biscuit period, one of the rats of the tartrate group died suddenly without preliminary indications of abnormality, leaving only 5 animals in this group for the remainder of the experiment.

TABLE XCIII

SUMMARY OF DATA, EXPERIMENT II

Figures are average weights (gm.) per rat of the particular groups

Periods	Kinds of biscuits			
	Phosphate A	Alum	Phosphate B	Tartrate B
End of fore (seventeen days). Biscuit	103	98	99	106
Twelfth day...............	102	105	109	118
Twenty-fifth day..........	126	112	134	141
Sixty-third day...........	149	123	155	158
Gain...................	46	25	56	52
Average daily amount (gm.) of food eaten				
Biscuit period..............	13.3	11.1	14.9	14.9

There were occasional instances of a mild degree of diarrhea in the rats of the alum group, and the experimenters regarded the animals of this group as less active and vigorous. They showed no symptoms of acute toxicity.

Experiment III. This was a study of the fecundity and of the normality of the offspring of the rats of Experiment II, the rats continuing on the biscuit diet (cheese and cabbage on Sunday) excepting as specified.

TABLE XCIV

SUMMARY OF DATA, EXPERIMENT III

Average body weights (gm.) until the occurrence of pregnancy

Period	Phosphate A	Alum	Phosphate B	Tartrate B
Fourth day.................	144	124	149	158
Eleventh day..............	149	121	158	163
Sixteenth day.............	154	121	153	164
Nineteenth day............	155	119	162	169
Average daily amount (gm.) of food eaten				
First twenty days..........	15.6	14.2	16.2	15.2

TABLE XCIV
EXPERIMENT III
FECUNDITY AND OFFSPRING

Period	Date of birth	Phosphate A No.	Wt.	Alum No.	Wt.	Phosphate B No.	Wt.	Tartrate B No.	Wt.
I	7/11	.....	..	.....	..	.....	..	8	75
	14	6[1]	67						
	17	.....	..	.....	..	11[2]	57		
						5[3]			
	18	8[4]	61	.....	..	8[5]			
	25	.....	..	.....	..	.....	..	10[6]	
	30	.....	..	.....	..	.....	..	9[7]	
	8/6	10	54						
II	12	.....	..	7[8]	52				
				3[9]	60				
III	9/12	11	33						
	13	8	46	9 D[10]	32	8	40		
	14	5	59	.....	..	.....	..	9	41
	.....		..	.....	..	.....	..	5	32
	15	.....	..	7	33	7	56		
	16	.....	..	.....	..	12	30		
	10/14	.....	..	D[11]					
Number litters............		6	..	4	..	6	..	5	
Total offspring............		48	..	26	..	51	..	41[12]	
Average number per litter..		8	..	6.5	..	8.5	..	8.2	
Lived one month..........		43[13]	..	24[14]	..	36[15]	..	31[16]	
Average weight at one month[17]............		.....	49	.....	39	.....	44	.....	54

[1] Two died a few hours after birth.   [2] One died 7/18; another 7/19.

[3] Each of the five rats, as it was born, was eaten by the mother.

[4] One died 7/18; two 7/19.

[5] One died 7/25; seven died 7/26, apparently killed by the mother.

[6] One died, 7/31.

[7] One died 7/30; seven eaten by mother 7/31.

[8] One died 8/12; the rest were unusually small but grew rapidly.

[9] One died 8/19; the rest were unusually small but grew rapidly.

[10] This prospective mother died 9/13, with nine half-grown embryos.

[11] The sixth rat of the alum group died 10/14, non-pregnant.

[12] There were only five females in this group because of the death of one of the original sextette on the second day of the biscuit period of the second experiment.

[13] Five died, some of them killed by mother.

[14] Two died.   [15] Fifteen died, some killed by mothers.

[16] Ten died, most killed by mothers.

[17] The surviving rats were normal; no appreciable difference, except that of size.

First Period (June 17 to August 6). The same adult male rats were associated from 11 P.M. to 9 A.M. in a cyclic order with each of the four groups of females on 12 different occasions. As soon as pregnancy had advanced to about the fourteenth day, the animal was removed from the group.

Second Period (August 7 to 22). Conditions the same but a new group of males substituted and associated with females continuously in regular rotation from cage to cage.

Third Period (August 23 to October 14). Conditions the same excepting biscuit diet discontinued and daily routine balanced diet fed.

During the first period, the male rats received the routine balanced diet; during the second and third period, they received the particular biscuit diet of the females with which they were associating at the time. Pregnant rats received the routine balanced diet. See Tables XCIII and XCIV.

TABLE XCV

SUMMARY OF DATA, EXPERIMENT IV

(Figures are average weights (gm.) per rat of the particular group.)

Periods	Phosphate A	Alum	Phosphate B	Tartrate B
End of fore................	88	85	83	87
Biscuit				
Fifth day...............	97	91	93	95
Twenty-eighth day........	121	108	126	132
Fifty-eighth day..........	170	138	171	180
Gain....................	82	53	88	93
After				
Seventh day..............	198	172	195	207
Fourteenth day........,...	210	183	200	218
Gain....................	40	45	29	38

Average daily amount (gm.) of food eaten

	Phosphate A	Alum	Phosphate B	Tartrate B
Fore.....................	21.8	21.2	21.0	21.2
Biscuit (inconclusive)*.......	16.2	15.7	17.4	17.3

* Not corrected for uneaten food which appeared more in the cage with the alum group.

The investigators reported no signs of acute toxicity, except with one of the rats of the alum group. On several occasions, separated by periods of apparent health, this animal exhibited varying degrees of paralysis of the hind legs, also diarrhea: it died October 10. At autopsy, there were no distinct indications of the cause of death except marked intestinal congestion. In view of the death of one rat early in the tartrate B period, it was impossible to be sure that death was not independent of the diet; yet what was believed to be a resemblance of the symptoms to aluminum poisoning made it impossible to contend that the diet was not responsible.

Experiment IV. In effect a duplication of the second but on male rats. The biscuits were leavened with the same baking powders and in the same amounts as in Experiment II and the same plan of feeding was followed. There were 8 males in each of four groups. There were no symptoms of acute poisoning. Slight diarrhea was present in the alum group, which animals appeared less active. See Table xcv.

"Summary and General Conclusions. We conducted four extended nutritional experiments on young healthy albino rats of both sexes, with biscuits of uniform composition but leavened with yeast, phosphate baking powders A and B, tartrate baking powders A and B, or alum baking powder (Success), under closely parallel conditions in all respects, in each experiment, and without material deviations or disparities of any character, except those introduced by the different kinds of biscuit under dietetic comparison. The results of these experiments present the following general conclusions:

"1. Rates of growth on the biscuits leavened with yeast, phosphate baking powders A or B or tartrate baking powders A or B, were so closely similar that the biscuits of these kinds were practically identical in nutritive value. The biscuits leavened with alum baking powder, however, were strikingly ineffective (in each experiment) in supporting growth, the extent of growth on the alum biscuits being (in general) only about 60 per cent of the amount of growth on the biscuits of the 5 other kinds named above.

"2. The relative nutritional ineffectiveness of the biscuits leavened with alum baking powder, as compared with the biscuits of the other 5 kinds named above, was due to direct toxic action of something in the biscuits, not to deficiency of nutrients in the biscuits, for the nutrients were practically the same in kind and proportion in all the biscuits.

"The 'something in the biscuits' leavened with alum baking powder, that caused the 'direct toxic action' referred to above was presumably (a) one or more soluble aluminum compounds, or (b) sulphate by-product, or both. That one or more soluble and absorbed aluminum compounds were responsible for some of the toxicity is indicated by the paralytic effects and other evidences of acute aluminum poisoning noted in the first and third experiments. That sulphate was responsible to some degree was suggested by the common occurrence of diarrhea in the rats (in each experiment) that ingested the alum biscuits.

"There were no symptoms of toxicity or diarrhea among the rats that received biscuits of the 5 other kinds named above.

"3. The diet of alum biscuits, compared directly with similar diets of biscuits leavened with phosphate (A), phosphate (B), or tartrate (B) baking powders, markedly interfered with reproduction. The diet of alum biscuits delayed the onset and unfavorably modified the outcome of pregnancy in, and impaired the nutrition of the offspring of, otherwise healthy and prolific albino rats.

"That a diet of alum biscuits, compared with diets of phosphate (A), phosphate (B), and tartrate (A) biscuits, impaired the fecundity of such rats was shown by the facts that (a) a smaller number (66 per cent) of the rats on the alum biscuit diet gave birth to litters; that (b) these rats bred later than those of the comparable groups on the three other kinds of biscuit; (c) gave birth to smaller litters and to smaller individuals in the litters; and, in one (the only) instance, died in pregnancy (with 9 half-grown embryos).

"That a diet of alum biscuits, compared with diets of phosphate (A), phosphate (B), and tartrate (B) biscuits impaired the nutrition of the offspring of otherwise healthy and prolific albino rats was shown by the facts that the offspring of the rats on alum biscuit diet (a) were smaller in number per litter; (b) smaller in size at birth; (c) smaller in size at the age of one month; and, (d) in one (the only) instance, were not developed to embryonic maturity.

"4. The results indicate, in general, that the two phosphate and the two tartrate baking powders used in these experiments are dietetically as acceptable for leavening purposes as yeast.

"5. Alum baking powders, if fairly represented by the Success brand (and we believe the samples of this brand used by us and purchased in the open marked are typical) are markedly inferior from the hygienic standpoint to well-known phosphate and tartrate varieties of baking powders.

"6. In their conclusion that 'alum baking powders are no more harmful than any other baking powders,' the Referee Board of Consulting Scientific Experts of the U. S. Department of Agriculture expressed an opinion (1914) that is not supported by the findings of these experiments."

Subsequent to the foregoing experimental studies, the same investigators conducted additional experiments. The following is an abstract of the report of their work.

The object of these experiments was to compare the relative nutritive qualities of alum baking powder and tartrate baking powder by determining the comparative nutritive effects of biscuits prepared with these powders on reproduction, on the fecundity of rats and on the normality of their offspring.

Animals and their environment. Two groups of 6 female animals were selected for each of 3 experiments, "from our own colony of albino rats, apparently normal in all respects." Each group contained representatives of several litters, but the number from each litter was the same in all the groups in an experiment.

Food for the animals. The diet for the rats from Monday to Saturday, inclusive, consisted exclusively of biscuits, except that on Thursday each rat was given 5 gm. of beef heart meat in addition. On Sunday cheese and cabbage (but no biscuits) were fed. More food of each kind than the animals could eat was offered daily. By presenting an excessive supply of biscuits six days a week, it was aimed to give the rats as much of this kind of food as could suitably be fed without stopping the growth of the animals on the "control" biscuits. The presentation of an abundance of cheese and cabbage on Sundays and of 5 gm. of meat per rat on Thursday served to reduce the effects of the unavoidable nutritive deficiency of an exclusive biscuit diet and to prolong the period of active growth.

For the preparation of alum biscuits they used, in the fifth experiment, the Success brand of alum baking powder (containing, according to its printed label, 30.75 per cent of sodium aluminum sulphate), and in the sixth and seventh experiments, the Parrot and Monkey brand (containing, according to its printed label, 32.67 per cent of sodium aluminum sulphate). The tartrate baking powder with which it was compared was one of the standard market brands.

The general recipe for the bakings was as follows:

Flour (4½ cups)...............................	502.2	gm.
Fat (4½ tablespoonfuls).......................	57.6	gm.
Salt (1½ teaspoonfuls)........................	8.2	gm.
Milk (¾ cup).................................	186.	gm.
Water..	187.	cc.
Baking-powder:		
Tartrate.............................	55.4904	gm.
Alum, "Success"............................	50.0850	gm.
Alum, "Parrot and Monkey"................	47.1150	gm.

Each experiment was conducted through a preliminary or "fore" period, during which the animals were given an abundance of the routine balanced laboratory diet. This food was presented on the following schedule: Sunday, fresh cabbage, cheese; Monday, cracked corn, beef heart; Tuesday, milk, wheat bread; Wednesday, fresh cabbage, beef heart; Thursday, fresh turnip, cheese; Friday, rye grain, wheat bread: Saturday, oats, beef heart.

Each group of animals received weighed amounts of food simultaneously, once a day (larger than the amounts they would probably eat); and the weights of the uneaten residues were recorded.

After the preliminary (fore) period, the feeding of biscuits was begun and continued until the rats gave birth to litters. Mother rats were fed the routine balanced diet to which was added a small portion of cheese.

Breeding. Two normal adult male rats were put into each of the cages containing the female rats at 11 P.M. and allowed to remain there until 9 A.M. the following day. At night they were put into cages with females, each group of males being put with the group of females other than that with which it had associated the day before. The procedure was continued in this cyclic way during the course of the breeding period. As soon as pregnancy advanced to about the fourteenth day in any rat, the animal was removed from the group and isolated in another cage, but the original males were associated with the remaining females in the groups on the cyclic schedule already referred to.

Experiment v. The experiment was divided into the following periods:

First (fore) period (May 4 to May 16). Rats fed the routine balanced laboratory diet.

Second period (May 17 to August 22). Biscuit fed; Sunday cheese and cabbage. On Thursdays beginning on the eighteenth day 5 gm. of meat per rat were fed in addition to biscuits. Between June 22 and July 28 breeding conditions were instituted, as described; breeding omitted on June 24 and July 8, 9 and 10, due to the hot weather. From July 13, two new groups of males (brothers of the original males) were substituted for the original males.

Third period (August 23 to September 22). The routine balanced diet was fed and the rats observed for pregnancy.

TABLE XCVI

SUMMARY OF DATA, EXPERIMENT V

AVERAGE WEIGHTS (GM.)

Periods	Kinds of biscuits	
	Tartrate	*Alum*
End of fore period..........................	81	84
Biscuit period—before breeding		
Tenth day (5/26)..........................	97	93
Twenty-third day (6/8).....................	120	107
Thirty-sixth day (6/21)....................	137	117
Gain......................................	56	33
Biscuit period (continued)—breeding period		
Fifty-first day (7/6)......................	152	121
Gain......................................	15	4
Total gain................................	71	37

The gain of the tartrate rats during the breeding period was undoubtedly due, in part, to the incidence of pregnancy.

The food eaten by the rats in the fifth experiment was indicated in the following table. The weights were obtained by subtracting the average weights of the uneaten portions of the food per day (daily food residues per rat) from the average weights of the food masses offered per day. Losses of water by evaporation from the food are unavoidably included in the values of food eaten in the calculation. Consequently the data in this table have a relative value only.

TABLE XCVII

FOOD EATEN BY RATS IN FIFTH EXPERIMENT

Date	Number of days	Kind of food eaten	Total amount of food eaten per rat in gm.		Average amount of food eaten per rat per day.		Average amount of food eaten per rat per day for all periods combined gm.	
			Tar.	Al.	Tar.	Al.	Tar.	Al.
5/4–16 (fore period)	13	Regular diet	.....	.....	.....	.....	......	......
5/17–6/21 (before breeding).	36	Biscuit, meat, cabbage, cheese	576	513	16.0	14.3	......	......
6/22–7/28 (breeding period).	37	Biscuit, meat, cabbage, cheese	587	470	15.9	12.7	......	......
7/29–8/22 (breeding discontinued).	25	Biscuit, meat, cabbage, cheese	395	341	15.8	13.6	......	......
8/23–9/22 (period of observation for pregnancy)......	16	Regular diet	.....	.....	.....	.....	......	......
Total No. of days food was weighed	98	....	1558	1324	....	....	15.9	13.5

The figures showed that the rats of the alum group ate less per day than the rats of the tartrate group during each period of the experiment. The investigators ascribed this difference to a cumulative toxic action of the alum biscuits. They recognized no reason to believe that a difference in palatability against the alum biscuits accounted for the effects observed.

TABLE XCVIII

SUMMARY OF DATA PERTAINING TO THE FECUNDITY AND OFFSPRING OF THE
RATS IN FIFTH EXPERIMENT

Diet	Tartrate	Alum
Number of rats that failed to bear litters........	1[1]	5[1]
Total number of offspring.....................	39	9
Average number of rats in a litter..............	7.8	9[2]
Total number of rats that lived for at least thirty days...	38	0
Average weight of surviving young rats at age of thirty days[3]..............................	44	[4]
Average weight of mothers thirty days after giving birth to litters.......................	164[5]	[6]

[1] These rats were given the routine balanced diet from August 23 to September 22 to ascertain whether the alum rats would catch up in weight with the tartrate rats. On September 22 the average weight of the tartrate rat was 179 gm. while that of the alum rat was 142 gm.
[2] This was the only litter in the alum group.
[3] The surviving rats were normal.
[4] All the young rats had died.
[5] Mother rats were normal.
[6] Mother rat died.

The investigators observed no signs of acute toxicity in the fifth experiment except in the case of the mother rat of the alum group. On August 10, six days after the birth of the litter, this rat had an abnormality which caused it to sag to the left side. The symptoms continued on August 11 and seemed about the same on August 12, but on the following day, August 13, they were unmistakably worse. Her condition was very bad on August 14. She rolled on her side and was unable to care for her young which were put up to her twice for nursing. The next morning, August 15, her condition was pitifully worse. Her respiration was very rapid and she was perfectly helpless. She seemed conscious of her young and tried to get to them. She endeavored to eat meat held to her mouth. Incessant rolling continued throughout the day and at 5 P.M. she was much worse. The next morning, August 16, she made a desperate effort to eat but was too paralyzed to swallow. She rolled from one end of the cage to the other again and again.

Two of her young died during the morning. At 10 A.M., August 17, the mother's eyes were set and glassy and 6 of the young were found dead. She was dead by noon and the last one of the litter died between 6 P.M. and 11 P.M.

The investigators concluded that the data showed clearly the following general effects on reproduction:

"1. The fecundity of the rats in the alum group was lower in degree than that of the rats in the tartrate group. Under uniform conditions for reproduction, except the nature of the two kinds of food,

"(a) Only 1 of the 6 rats of the alum group gave birth to a litter; this mother showed symptoms of acute toxicity and died eighteen days after giving birth to a litter;

"(b) Of the 6 rats of the tartrate group, 5 gave birth to litters; all the mothers survived in good condition.

"2. (a) All the offspring of the mother of the alum group died within eighteen days after being born;

"(b) Only 1 of a total number of 39 young rats in the tartrate group died; all the surviving rats were normal."

The investigators held that it was obvious from the foregoing data that the diet of alum biscuits, under the parallel conditions of these experiments and compared directly with biscuits leavened with tartrate baking powder, markedly interfered with the normal onset, progress, and outcome of pregnancy in otherwise healthy and prolific albino rats. Although all the offspring of the mother that was fed alum biscuits died, no statement could be made regarding the effect of these biscuits on the nutrition of the off-spring, except that the rats were very small at time of birth, because the mother rat died eighteen days after giving birth to litter and the young rats were therefore uncared for. It is possible that the alum biscuits had so affected her that she succumbed under the strain of caring for her young.

Experiment VI. The sixth experiment was, in effect, a duplicate of the fifth. The experiment was divided into the following periods:

First (fore) period (June 30 to July 7). Rats fed the routine balanced laboratory diet.

Second period (July 8 to September 22). Biscuits fed; Sunday, cheese and cabbage. On Thursday, beginning on the eighth day, 5 gm. of meat per rat were fed in addition to biscuits. Breeding conditions, as described, were instituted from August 2 to August 11 and from August 22 to August 31.

Third period (September 23 to October 22). The routine balanced diet was fed and the rats observed for pregnancy.

TABLE XCIX

SUMMARY OF DATA, EXPERIMENT VI

AVERAGE WEIGHTS (GM.)

Period	Kinds of biscuit	
End of fore period	Tartrate 110	Alum 109
Biscuit period—before breeding		
Eleventh day (July 18)....................	128	119
Twenty-fifth day (August 1).................	144	132
Gain.......................................	34	23
Biscuit period (continued)—breeding period		
Thirty-fifth day (August 11)................	156	134
Gain.......................................	12	2
Biscuit period (continued)—breeding discontinued		
Forty-third day (August 19).................	174	145
Gain.......................................	18	11
Total gain.................................	64	36

The data showed that the gain in weight was less for the rats on the alum biscuits than for those on the tartrate biscuits. This was in accord with the results of the fifth experiment. The greater gain of the tartrate rats was undoubtedly due in part to the incidence of pregnancy.

TABLE C

FOOD EATEN BY RATS IN SIXTH EXPERIMENT

Date	No. of days	Kind of food eaten	Total amount of food eaten per rat, gm.		Average amount of food eaten per rat per day, gm.		Average amount of food eaten per rat per day for all periods combined, gm.	
			Tar.	Al.	Tar.	Al.	Tar.	Al.
6/30–7/1 (fore period)	8	Regular diet						
7/8–8/1 (before breeding).	25	Biscuit, meat, cabbage, cheese	378	336	15.1	13.4		
8/2–8/11 (breeding period).	10	Biscuit, meat, cabbage, cheese	135	108	13.5	10.8		
8/12–21 (breeding discontinued.)	10	Biscuit, meat, cabbage, cheese	149	115	14.9	11.5		
8/22–31 (breeding resumed).	10	Biscuit, meat, cabbage, cheese	216	128	21.6	12.8		
9/1–8 (breeding discontinued.) 9/1–22	8 22	Biscuit, meat, cabbage, cheese	190	 397	23.8	18.0		
9/23–10/22 (period of observation for pregnancy)	29	Regular diet						
Total No. of days food was weighed Tartrate Alum	63 77	....	1068	 1084	....	....	17.0	14.1

The figures showed that the rats of the alum group ate less per day than the rats of the tartrate group during every period of the experiment that food weights were made. As in the first experiment the investigators ascribed this to a cumulative toxic action of the alum biscuits.

TABLE CI

SUMMARY OF DATA PERTAINING TO THE FECUNDITY AND OFFSPRING OF THE
RATS IN THE SIXTH EXPERIMENT

Diet	Tartrate	Alum
Number of rats that failed to bear litters..........	0	2
Total number of offspring.......................	45	22
Average number of rats in litter.................	7.5	5.5
Total number of rats that lived for at least 28 days..	44	18
Average weight of surviving young rats at age of 28 days,[1] grams................................	38	35
Average weight of mother immediately following birth of litter, gm...........................	177	145
Average weight of mother 28 days after giving birth to litter,[2] grams.............................	163	143[3]

[1] The surviving rats appeared to be normal.
[2] The mothers appeared to be normal.
[3] One mother of the alum group was removed from the experiment the fifth day after giving birth to 1 rat. The young rat died three days after birth.

There were no signs of acute toxicity in the sixth experiment. The investigators concluded that the data showed clearly the following general effects on reproduction:

"1. The fecundity of the rats in the alum group was lower in degree than that of the rats in the tartrate group. Under uniform conditions for reproduction, except the nature of the two kinds of food,

"(a) Only 4 of the 6 rats of the alum group gave birth to litters: 1 of the litters consisted of only 1 rat;

"(b) All the rats of the tartrate group bred normally.

"2. The offspring of the rats of the alum group were inferior to those of the rats in the tartrate group in being

"(a) smaller in size at birth;

"(b) smaller in number per litter;

"(c) smaller in size at age of twenty-eight days.

"3. (a) Four of a total number of 22 young rats in the alum group died within twenty-eight days:

"(b) Only 1 of a total number of 45 young rats in the tartrate group died within twenty-eight days.

"4. The average weight of mother rats immediately following and twenty-eight days after birth of litters was less for the rats fed alum biscuits than for those on the tartrate biscuit diet.

"The results of this experiment are in accordance with those obtained in the fifth experiment. The alum biscuits, compared directly with biscuits leavened with tartrate baking powder, interfered with the normal onset, progress, and outcome of pregnancy in, and noticeably impaired the nutrition of, the offspring of otherwise healthy and prolific albino rats."

Experiment VII. The seventh experiment was practically a duplicate of the fifth and sixth experiments. The rats, 6 females in each of 2 groups, were obtained from 2 litters.

Their ages on the first day of the recorded fore period ranged from twenty-seven to thirty-three days.

All the rats (2 in each group) whose age was thirty-three days were removed from the experiment during the fifteenth day of the fore period (June 21) and replaced by other rats forty-eight days old, because they were found to be abnormal.

The experiment was divided into the following periods:

First (fore) period (June 7 to June 28). Rats fed the routine balanced laboratory diet.

Second period (June 29 to October 1). Biscuits fed; Sunday, cheese and cabbage. On Thursdays, 5 gm. of meat per rat were fed in addition to biscuits. Breeding conditions, as described, were instituted from August 12 to August 21, and from September 1 to September 10. The 2 groups of male rats used for breeding were from the sixth experiment and were used during the intervals when breeding was discontinued in that experiment.

TABLE CII
SUMMARY OF DATA, EXPERIMENT VII
AVERAGE WEIGHTS (GM.)

Periods	Kinds of biscuit	
	Tartrate	Alum
End of fore period.............................	74	75
Biscuit period—before breeding		
Tenth day (July 8)..........................	86	84
Twenty-third day (July 21)..................	100	96
Forty-fourth day (August 11)................	131	113
Gain...	57	38
Biscuit period (continued) breeding period		
Fifty-third day (August 20)..................	139	118
Gain...	8	5
Biscuit period (continued)—breeding discontinued:		
Sixty-third day (August 30)..................	162	134
Gain...	23	16
Total gain...................................	88	59

The data showed, in accord with the results obtained in the fifth and sixth experiments, that the gain in weight was less for the rats on the alum biscuits than for those on the tartrate biscuits. As in the previous experiments the difference in weight might be accounted for, in part, by the incidence of pregnancy.

TABLE CIII

FOOD EATEN BY RATS IN THE SEVENTH EXPERIMENT

Date	No. of days	Kind of food eaten	Total amount of food eaten per rat, gm.		Average amount of food eaten per rat per day, gm.		Average amount of food eaten per rat per day for all periods combined, gm.	
			Tar.	Al.	Tar.	Al.	Tar.	Al.
6/7–28 (fore period)............	22	Regular diet						
6/29–8/11 (before breeding).	44	Biscuit, meat, cabbage, cheese	599	562	13.7	12.8		
8/12–21 (breeding period).	10	Biscuit, meat, cabbage, cheese	130	111	13.0	11.1		
8/22–31 (breeding discontinued).	10	Biscuit, meat, cabbage, cheese	187	160	18.7	16.0		
9/1–10 (breeding resumed).	10	Biscuit, meat, cabbage, cheese	150	132	15.0	13.2		
9/11–20 (breeding discontinued).	10	Biscuit, meat, cabbage, cheese	184	181	18.4	18.1		
Total no. of days food was weighed	84	....	1250	1146	....	....	14.9	13.7

It was evident from the figures in this table that the rats of the alum group ate less than the rats of the tartrate group during every period of the experiment that food weights were made.

TABLE CIV

SUMMARY OF DATA PERTAINING TO THE FECUNDITY AND OFFSPRING OF THE
RATS IN THE SEVENTH EXPERIMENT

Diet	Tartrate	Alum
Number of rats that failed to bear litters...........	0	0
Total number of offspring.......................	50	40
Average number of rats in a litter...............	8.3	6.7
Total number of rats that lived for at least twenty-eight days[1]..................................	47	27
Average weight of surviving rats at age of twenty-eight days, gm................................	33	45
Average weight of mother immediately following birth of litter, gm...........................	149	130
Average weight of mother twenty-eight days after giving birth to litter,[2] gm.....................	143	151

[1] The surviving rats appeared to be normal.
[2] The mothers appeared to be normal.

There were no signs of acute toxicity in the seventh experiment. The investigators concluded that the data showed the following general effects on reproduction:

"1. The fecundity of the rats was the same in each group; all the rats of each group gave birth to litters. This result differs from that obtained in the fifth and sixth experiments. However, the rats in the alum group bred later than those in the tartrate group.

"2. The offspring of the rats of the alum group were inferior to those of the rats in the tartrate group in being

"(a) smaller in size at birth;

"(b) smaller in number per litter.

"The average weight of surviving young rats at twenty-eight days of age was greater in the alum group than in the tartrate group. This result is not in accordance with that obtained in the sixth experiment and we believe that this result may be accounted for in part by the smallness of the litters which the alum group rats had to raise. Data of this nature were not obtained in the fifth experiment since all the young rats died within thirteen days after birth.

"3. (a) Thirteen of a total number of 40 young rats in the alum group died within twenty-eight days;

"(b) Three of a total number of 50 young rats in the tartrate group died within twenty-eight days.

"4. The average weight of mother rats immediately following the birth of litters was less for the rats fed alum biscuits than for those on the tartrate biscuit diet. The reverse was the case however regarding the weight twenty-eight days after the birth of litters. This may be accounted for in part by the fact that the alum biscuit mothers had fewer young rats to nourish and were therefore not subjected to as great a strain as were the rats on tartrate biscuit diet.

"The results of this experiment are, in the main, in accordance with those obtained in the fifth and sixth experiments.

"*Summary of General Conclusions.** We conducted 3 nutritional experiments on healthy albino rats, with biscuits of uniform composition but leavened respectively with alum and tartrate baking powders under closely parallel conditions in all respects, in each experiment and without material deviations or disparaties of any character except those introduced by the different kinds of biscuit under dietetic comparison. The results of these experiments present the following general conclusions:

"1. Rates of growth on the biscuits leavened with alum baking powder were less than those on biscuits leavened with tartrate baking powder.

"2. The diet of alum biscuits, compared directly with a similar diet of biscuits leavened with tartrate baking powder, markedly interfered with reproduction. The diet of alum biscuits delayed the onset (in two of three experiments) and unfavorably modified (in two of three experiments) the outcome of pregnancy in, and impaired the nutrition (in all the experiments) of, otherwise healthy and prolific albino rats.

"That the diet of alum biscuits, compared with tartrate biscuits, impaired the fecundity of rats was shown by the facts that (a) a smaller number of rats on the alum biscuit diet gave birth to litters (in two experiments); that (b) these rats bred later than those on the tartrate biscuit diet (in two experiments); that (c) they gave birth to smaller litters and to smaller individuals in the litters (in all the experiments).

"That the diet of alum biscuits, compared with tartrate biscuits, impaired the nutrition of the offspring of otherwise healthy

* Experiments v, vi and vii.

and prolific albino rats was shown by the facts that the offspring of the rats on the alum biscuit diet (a) were smaller in number per litter; (b) smaller in size at birth; and (c) that they died in greater numbers before reaching the age of twenty-eight to thirty days.

"3. The average weight of surviving young rats at the age of from twenty-eight to thirty days was less in the alum group than in the tartrate group in one experiment but more in another experiment. In one experiment no data of this nature were obtained because all the young rats in the alum group (there was only one litter in this group) died within thirteen days after birth. The greater weight of the rats in the alum group in one experiment may be accounted for in part by the smallness of the litters which the alum group rats had to raise.

"4. The average weight of mother rats immediately following birth of rats was less for the rats on the alum biscuit diet than for those on the tartrate biscuit diet (these weights were made in two experiments).

"5. The average weight of mother rats twenty-eight to thirty days after birth of litters was less in the alum group than in the tartrate group in one experiment but more in another experiment. In one experiment no data of this nature were obtained because the only mother in the alum group died thirteen days after giving birth to litter. The greater weight of the mothers in the alum group in one experiment may be accounted for in part by the fact that these mothers had fewer young rats to nourish and were therefore not subjected to as great a strain as were the rats of the biscuit diet.

"6. The relative nutritional ineffectiveness of the biscuits leavened with alum baking powder as compared with biscuits leavened with tartrate baking powder was due to direct toxic action of something in the biscuits, not to deficiency of nutrients in the biscuits, for these were the same in all the biscuits."

Philip B. Hawk, Clarence A. Smith, Olaf Bergeim and Earl A. Shrader[2] conducted feeding experiments with albino rats to study growth and reproduction. These were practically repetitions of the experiments conducted by Gies and coworkers (pp. 250–271).

There were three series of experiments, 12 animals being used in each series, divided into groups of 6 each, to one of which food prepared with a so-called alum baking powder (Success brand) was fed and, to the other of which food prepared with a so-called tartrate baking powder (Royal brand) was fed. Diet "A" was identical with the "routine balanced laboratory diet" of Gies. Diet "B" consisted of fresh cabbage and cheese, on Sunday; meat and Royal biscuits on Wednesdays; and Royal biscuits only at all other times. Diet "C" was the same as diet "B," except that Success biscuits were fed in place of Royal biscuits. The biscuits were prepared by the same recipe as was employed by Gies (p. 250), 3 level teaspoonfuls of the respective powders to one cup of flour being used. Care in the selection of animals and satisfactory laboratory conditions were maintained.

Periods: Fore, days 1 to 7; Biscuit, days 8 to 77; Breeding, days 78 to 182.

TABLE CV

SUMMARY OF WEIGHTS, GM.

Groups	Alum			Tartrate		
	I	IA	IB	2	2A	2B
Day 7, weight of group...............	270	216	264	216	265	266
weight, average...............	45	36	44	44	36	44
Day 77, weight of group..............	744	637	704	743	808	764
weight, average..............	124	106	117	135	124	127
Gain, weight of group..............	474	421	440	527	543	498
weight, average...............	79	70	73	91	88	83

Gain in weight, total for 18 alum rats...................... 1395
Total for 18 tartrate rats................................. 1568
Average for alum rats.................................... 74
Average for tartrate rats................................. 87.5

Conclusion: "The ingestion of bread made with tartrate baking powder therefore apparently caused more rapid growth of the young rats than did the ingestion of similar bread made with alum powder."

In the mating process the male animals were associated with the females in rotation, much as described in the experiments by Gies.

TABLE CVI
NUMBER OF PREGNANCIES IN ONE-HUNDRED AND FIVE DAYS

Group	Alum			Tartrate		
	I	I A	I B	2	2A	2B
	3	2	3	4	4	2

TABLE CVII
FECUNDITY AND OFFSPRING
ALUM

Time of birth days after 1st mating	Number rats born	Died	Weights at end of thirty days, gm.	
			Total of litters	Average rat
30	7	......	273	39
42	10	....	340	34
49	7	....	266	38
35	4	....	164	41
36	3	....	129	43
42	9	....	324	36
44	5	....	200	40
49	9	4	175	35

TARTRATE

29	8	....	344	43
34	5	....	260	52
43	8	2	252	42
41	13	1	384	32
29	8	....	312	39
29	8	....	336	42
34	6	1	245	49
43	5	5 eaten	...	...
34	6	1	235	47
40	9	2	308	44

The average length of time elapsing between the first mating and the birth of the litters was forty days for the alum groups and thirty-five days for the tartrate groups.

Not only were more litters born to the mothers in the tartrate group, but the total number of offspring was greater. The alum rats gave birth to a total of fifty-four young, whereas the tartrate animals gave birth to 76 offspring. The fecundity of the tartrate animals was therefore over 40 per cent greater than that of the alum rats.

The young rats born to the tartrate animals also grew somewhat more rapidly than did those born to the mothers of the alum group.

"*General Conclusions.* 1. Albino rats fed on alum baking powder biscuits showed a definite retardation of growth as compared with rats fed on biscuits made with tartrate baking powder.

"2. Fewer of the albino rats fed on alum biscuits became pregnant and smaller litters were obtained on the average. Pregnancy also developed later in the alum group.

"3. No marked differences were noted in the rates of growth of the young from mothers fed on the two types of biscuits."

THE AUTHOR'S ANALYSIS OF DATA RELATING TO THE FOREGOING DIETS[2]

Using the food value of flour based on all analyses and of milk in Bulletin 28, U. S. Department Agriculture, "Composition of American Food Material," viz., for flour: water 11.4 per cent, protein 10.6 per cent, fat 1.1 per cent, carbohydrates 76.3 per cent, ash 0.6 per cent, and for milk: water 87.0 per cent, protein 3.3 per cent, fat 4.0 per cent, carbohydrates 5.0 per cent, ash 0.7 per cent; and assuming the s.a.s. baking powder to yield 13 per cent CO_2 and contain the corresponding amount of aluminum sulphate together with a total sulphate corresponding to 39.51 per cent Na_2SO_4, as determined by the writer on the "Success" sample supplied by Professor Gies, the composition of the s.a.s. baking powder biscuits, as dry solids, calculated by the writer, was:

TABLE CVIII

Per Cent

Gies Experiment 1.
Baking powder residue

Aluminum oxide Al_2O_3	0.30 } 2.7
Sodium sulphate Na_2SO_4	2.39 }
Salt	1.5
Ash of flour and milk	0.8
Protein	10.9
Fat	5.8
Carbohydrate	78.3

TABLE CVIII. (*Continued*)

Per Cent

Gies Experiments 2 to 7, inclusive.
Baking powder residue

Aluminum oxide Al_2O_3	0.43	3.8
Sodium sulphate Na_2SO_4	3.4	
Salt		1.4
Ash of flour and milk		0.7
Protein		10.3
Fat		12.1
Carbohydrate		71.6

For the tartrate biscuits, assuming as above and that the tartrate baking powder yielded 13 per cent CO_2 and contained total tartaric acid to the amount of 29.09 per cent, as found by the writer in a sample of Royal baking powder, the composition of the tartrate baking powder biscuits, as dry solids, was:

TABLE CIX

Per Cent

Gies Experiments 2 to 7, inclusive.
Baking powder residue

Rochelle salt, $NaKC_4H_4O_6$	3.9
Salt	1.4
Ash of flour and milk	0.7
Protein	10.2
Fat	12.0
Carbohydrate	71.7

The directions for making biscuits with Royal baking powder, as given on the can (and these are the common directions for tartrate baking powders), stated:

Sift together 4 cups of flour, 2 rounded teaspoonfuls Royal baking powder and ½ teaspoonful salt. Rub in 2 tablespoonfuls shortening.

By trial, the writer found that 1 rounded teaspoonful of Royal baking powder weighed 8.87 gm. This corresponded to 4.435 gm. Royal baking powder to each cup of flour. The amount used by Gies in Experiments 2 to 7, viz., 12.3312 gm., was thus 2.76 times that advocated for the housewife's use in making biscuit.

The directions for making biscuits with Success baking powder, as given on the can (and these are the common directions for straight s.a.s. baking powders), stated:

Into one quart of sifted flour put *one* rounded teaspoonful of the powder; mix thoroughly by sifting and add one tablespoonful of shortening.

By trial, the writer found that one rounded teaspoonful of Success baking powder weighed 7.577 gm. This corresponded to 1.894 gm. Success baking powder to each cup of flour. The amount used by Gies in Experiments 2 to 7, viz., 11.0217 gm., was thus 5.82 times that advocated for use by the housewife in making biscuit. The reason for advocating the use of a smaller quantity of an s.a.s. baking powder as compared with a tartrate baking powder is due to the fact that the rate of evolution of the gas of an s.a.s. powder is slower and hence its efficiency for a given amount of gas is greater. It thus appears that Gies' tartrate baking powder biscuit contained a residue of Rochelle salts 2.78 times what would result from the quantity advised by the manufacturers in the use of the powder and that Gies' s.a.s. baking powder biscuit contained a baking powder residue 5.82 times what would result from the quantity advised by the manufacturers in the use of the powder. It is to be kept in mind that it is on this basis that Gies made his comparative experiments.

Regarding the nutritive value of Gies' biscuits as exclusive diets for growing albino rats, they presented the following defects, criticisms applicable alike to the tartrate and s.a.s. baking powder diets:

Proteins. The *quality* of the protein was 90 per cent wheat, 10 per cent milk. It is extremely doubtful whether the amino-acid deficiencies of the wheat protein would be overcome by the 10 per cent milk proteins.

The *quantity* of the total protein, viz., 10.2 per cent and 10.3 per cent, was unquestionably below what will maintain and promote normal growth of the albino rat. Fifteen per cent is a desirable minimum in any experiments where the protein quantitative factor is to be regarded as adequate.

Mineral Constituents. The wheat mineral ingredients are lacking in the mineral constituents essential for maintenance and growth. The addition of 0.2 per cent milk salines to the dietary is insufficient to overcome the wheat mineral deficiency both as to quality and quantity.

Vitamines. Vitamines are so widely distributed and their avoidance, where there is any variation in the food supply, so difficult, that one hesitates to state with positiveness that any given dietary was lacking in vitamines. Nevertheless, considering the character of these biscuit diets, the opinion is formed that they were so low in vitamines as to be decidedly deficient. Professor Gies apparently recognized this defect when he fed the animals on their special diets only on week days; and on Sundays gave the general diet of the laboratory animals, to supply any vitamine defect. It is not at all convincing, however, that any decided vitamine deficiency for six days would be compensated for by a general diet on the seventh. Particularly is this so if the condition of the animals was poor from other food deficiencies.

It thus appears that Gies' experiments were conducted with diets nutritionally deficient, namely (1) in protein quality and quantity, (2) in mineral salts and (3) in vitamines, which deficiencies he attempted to correct by a general diet on one day a week; and that although the amount of baking powder residue was substantially the same in both the tartrate and the s.a.s. baking powder diets, yet this was 2.78 times that which would result from the normal use of the tartrate baking powder as directed, contrasted with 5.82 times as much in the case of the s.a.s. baking powder.

EXPERIMENTS BY E. E. SMITH[2]

The following feeding experiments were conducted to parallel the experiments of Gies, Heft and Karshan, herein after referred to as Gies' experiments, under conditions of a nutritionally adequate diet.

Group 1. Five albino rats, 1 male and 4 females; preliminary observation thirty-seven days; fed a "standard" diet for a period of one hundred and fifty-four days.

Group 2. Five albino rats, 2 males and 3 females; preliminary observation thirty-seven days; fed a standard diet to which was added Rochelle salt, for a period of one hundred and fifty-four days.

Group 3. Five albino rats, 2 males and 3 females; preliminary observation thirty-seven days; fed a standard diet to which was added sodium sulphate, for a period of one hundred and fifty-four days.

During the preliminary period of thirty-seven days all animals were fed a general mixed diet to determine normality of health and growth.

Food Preparations. 1. Biscuits raised with hydrochloric acid exactly neutralizing sodium bicarbonate, so as to yield a residue of sodium chloride in known amount.

Recipe	Gm.
Patent wheat flour	600
Soy bean flour	200
Sodium bicarbonate ⎰ yielding NaCl	
Dilute hydrochloric acid ⎱	12
Water	377

The dry ingredients were thoroughly mixed by double sifting into a bread mixer, the wet ingredients added and the whole quickly but thoroughly mixed, transferred in biscuit-size portions to greased pans and baked, as in common domestic practice. The biscuits while still hot were broken open and subsequently air dried and comminuted so as to pass through a 16 by 18 to the inch mesh sieve. Thus prepared, it constituted the "standard biscuit powder" of the various diets indicated.

2. "Mixed Salts." Prepared as described by Osborne and Mendel.[5]

3. Butter. This was sweet butter of high grade, as purchased.

The compositions of these constituents of the diet were calculated as in the case of the food of the Gies experiments, excepting where actual determinations were made, as reported. Determinations of composition of standard biscuit powder; moisture 8.48 per cent; protein 21.14 per cent of dry solids; ash 3.01 per cent of dry solids.

TABLE CX

CALCULATED COMPOSITION OF STANDARD BISCUIT POWDER AS DRY SOLIDS

	Per Cent
"Baking powder residue," NaCl	1.7
Ash of flour	1.3
Protein	21.1
Fat	6.4
Carbohydrate	69.5

TABLE CXI

DIET MIXTURES AS FED AND COMPOSITIONS

Group 1. Standard.

Mixture

	Gm.
Standard biscuit powder	160
Mixed salts	8
Butter	32

Dry solids

Composition	Per Cent
"Baking powder residue," NaCl	1.36
Mixed salts	4.37
Ash of flour	1.04
Protein	16.9
Fat	20.5
Carbohydrate	55.7

Group 2. Rochelle Salt

Mixture	1 Gm.	2 Gm.
Standard biscuit powder	149.74	144.61
Rochelle salt	10.26	15.39
Mixed salts	8	8
Butter	32	32

Composition Dry solids	1 Per Cent	2 Per Cent
"Baking powder residue," NaCl	1.29	1.25
Rochelle salt	4.22	6.38
Mixed salts	4.42	4.44
Ash of flour	0.98	0.95
Protein	16.0	15.5
Fat	20.3	20.3
Carbohydrate	52.6	51.0

TABLE CXI. (*Continued*)

Group 3. *Sodium Sulphate*

Mixture	1 Gm.	2 Gm.
Standard biscuit powder	152.5	148.75
Sodium sulphate (Na_2SO_4)	7.5	11.25
Mixed salts	8.0	8.0
Butter	32.0	32.0

Composition	Dry solids	1 Per Cent	2 Per Cent
"Baking powder residue," NaCl		1.39	1.26
Sodium sulphate		3.67	5.99
Mixed salts		4.38	4.36
Ash of flour		0.99	0.96
Protein		16.1	15.6
Fat		20.2	20.0
Carbohydrate		53.1	51.5

In planning the above experimental observations, it was thought possible that in the Gies experiments the diarrhea observed with the s.a.s. baking powder group might be due to the laxative action of the large amount of sodium sulphate residue from the excessive use of s.a.s. baking powder in the s.a.s. baking powder biscuit. No such laxative action was obtained. To observe the effect of the s.a.s. baking powder used in reasonable amount, a group of 5 rats had been started on an s.a.s. baking powder biscuit diet, in which the amount of powder used was not unreasonably excessive. In the absence of laxative action of the sulphate group, the amount of s.a.s. baking powder in the food of this group was increased. The data are:

Group 4. Five albino rats, males, fed on straight s.a.s. baking powder biscuit for a period of one hundred and forty-seven days.

Recipe of baking powder biscuit:

	1		2	
Patent wheat flour	600	gm.	600	gm.
Soy bean flour	200	gm.	200	gm.
"Success" baking powder	15.14	gm.	79.0	gm.
Salt (sodium chloride)	12	gm.	12	gm.
Water	456	cc.	456	cc.

The process was the same as that employed for making standard biscuit powder, and the product will be designated, "Baking Powder Biscuit Powder."

TABLE CXII

CALCULATED COMPOSITION OF BAKING POWDER BISCUIT POWDER
Dry solids

	1 Per Cent	2 Per Cent
Baking powder residue		
Aluminum oxide (Al_2O_3)	0.10	0.50
Sodium sulphate (Na_2SO_4)	0.81	3.94
Added Salt	1.63	1.52
Ash of flour	1.65	1.54
Protein	18.7	17.2
Fat	6.3	5.9
Carbohydrate	70.8	69.2

TABLE CXIII

DIET MIXTURE, AS FED AND COMPOSITION

Group 4. Baking Powder	1	2
Mixture	Gm.	Gm.
Baking powder biscuit powder, #1	160	
Baking powder biscuit powder, #2	...	160
Mixed salts	8	8
Butter	32	32

Composition Dry solids	1 Per Cent	2 Per Cent
Baking powder residue		
Aluminum oxide (Al_2O_3)	0.08 ⎰ 0.73	0.41 ⎰ 3.57
Sodium sulphate (Na_2SO_4)	0.65 ⎱	3.16 ⎱
Mixed salt	4.33	4.33
Added salt	1.33	1.24
Ash of flour	1.31	1.22
Protein	17.8	16.3
Fat	20.2	19.9
Carbohydrate	54.3	53.6

In respect to the above diets of the writer's experimental observations, hereinafter designated the Smith experiments, it will be observed:

That regarding the *quality* of the *protein* ingredients, the combination of soy bean flour with the patent wheat flour provided in a substantial way such variety of the amino-acid components of the proteins of the diet that any deficiency in quality of such components was entirely avoided. Moreover, the *quantity* of total protein was between 15.5 per cent and 17.8 per cent, an amount entirely adequate to supply protein in an exclusive diet of the foods in question.

That regarding the *mineral ingredients* of the various diets, the addition of the mixed salts to the diets in amounts varying from 4.33 to 4.44 per cent of the dry components entirely avoided any deficiency in mineral ingredients of the diets. Indeed, in view of the fact that there was present in all the diets, either from added salt or from baking powder residues, sodium chloride varying from 1.22 per cent to 1.36 per cent, it may well be questioned whether the chloride component of the mixed salts did not make the amount of total chloride in the diets excessive. This possibility cannot be denied. It is certain, however, that the mixed salts did avoid any mineral deficiencies in the diet, either as to quality or quantity of the mineral ingredients.

That regarding *vitamines* in the diets, the use of soy bean flour provided an abundance of water-soluble vitamines and the use of sweet butter in the amount present provided an abundance of fat-soluble vitamine. As these are the vitamines essential to rat maintenance and growth, it seems clear that any vitamine deficiency was avoided by the diets in question.

A comparison, then, of the above diets with those of Gies seems to show clearly that the diets of the two series of investigations differ in this important respect: that the Gies diets were deficient in certain ingredients, namely, in quality and quantity of protein, the quantity and quality of the mineral ingredients and in the vitamines; and that the Smith diets showed none of these deficiencies. Further, it should be considered that Gies relied upon a general diet once a week and that Smith fed his diets continuously throughout the period of investigation.

The proposition of overcoming a deficient diet by a general diet one day a week, when based upon superficial consideration of human individuals, seems logical enough. However, this conclusion loses sight of actual conditions that pertain with a small animal like the rat. One of the very reasons for its use in experimental work is the fact that the permanent effects of any dietary become apparent in a relatively short time. Thus, the ultimate effects of complete starvation results in four or five days with rats; while with larger animals, such as the dog, and with humans the fatal outcome may be delayed for many weeks. The period of complete growth to maturity of the white albino rat is about two-hundred and eighty days. If we regard eighteen years as the period of complete growth to maturity of the human individual, then the proportion of that time corresponding to one week of the rat's period of growth would be six months, and to limit the rat for six days to a deficient diet expecting it to overcome the deficiency by a general diet on the seventh day and thus to continue for sixty days would correspond to limiting a human growing individual to a deficient diet for 22.3 weeks and expect it to overcome the deficiency by a general diet for 3.7 weeks, repeating this for four years or longer. It is fair to believe, that few would expect a child thus fed from the age of two or three years through the succeeding four years of its life to be maintained in health and to exhibit normal growth during such period. It is also unfair to assume that the rat fed a deficient diet for six days and allowed a general diet for one day, over a period of sixty days, would be maintained in health or exhibit normal growth. The addition of meat on Thursdays in the later experiments and in the Hawk and C. A. Smith experiments would do little to correct this. Hence, we must conclude that the Gies experiments were with animals on a deficient diet, even though they were allowed a general diet one day out of every seven.

We have already stated that the plan of the Smith experiments had in view the possibility that in the Gies experiments the diarrhea of the s.a.s. baking powder group was due to the laxative action of the sodium sulphate from the excessive quantity of the baking powder used in the preparation of their food, even though the Rochelle salt of the corresponding group did not produce such laxative action. The outcome of the Smith experiments did not support this view, as will be apparent from the following results. All diets were fed *ad libitum*.

TABLE CXIV

Group 1.   Standard
Preliminary period on general diet.............................. 37 days
Period on standard diet...................................... 154 days
Results:
  No illness
  No diarrhea
  Growth: 1 male, 99 per cent of curve of Osborne and Mendel
          4 females, 89 per cent curve of Osborne and Mendel
          Average 91 per cent curve of Osborne and Mendel

Group 2.   Rochelle Salt
Preliminary period on general diet.............................. 37 days
Period 1.   Rochelle salt diet 1................................ 28 days
Period 2.   Rochelle salt diet 2................................ 126 days
Results:
  No illness
  No diarrhea
  Growth: End of Period 1.   90 per cent of standard
          End of Period 2.   86 per cent of standard

Group 3.   Sodium Sulphate
Preliminary period on general diet.............................. 37 days
Period 1.   Sodium sulphate diet 2............................. 28 days
Period 2.   Sodium sulphate diet 2............................. 126 days
Results:
  No illness
  No diarrhea
  Growth: End of Period 1.   97 per cent of standard
          End of Period 2.   86 per cent of standard

Group 4.   s.a.s. Baking Powder
Preliminary period on general diet.............................. 44 days
Period 1.   s.a.s. baking powder diet 1........................ 42 days
Period 2.   s.a.s. baking powder diet 2........................ 105 days
Results:
  No illness
  No diarrhea
  Growth: End of Period 1.   98 per cent of standard
          End of Period 2.   81 per cent of standard

The slightly lower weight curves of Group 1, as compared with the normal weight curves of Osborne and Mendel may be due either to the character of the standard diet or to the particular breed of albino rats.

COMPARISON OF RESULTS

The results show no illness or diarrhea from Rochelle salts, sodium sulphate or s.a.s. baking powder residue, even in the excessive amounts or for the prolonged periods of experimental observation, viz., about two and one-half times the periods of observation in Experiments 1, 2 and 4 of Gies. This absence of illness is in direct variance with Gies' results, even though there was no intermission for one day a week, as was the case with Gies' animals, which on Sunday were given a general laboratory diet. It will be recalled that Gies found no illness with the Rochelle salt group but found manifestations with the s.a.s. baking powder biscuit which he attributed to "direct toxic action" of the baking powder residue of such biscuits. The amount of such residue was 32 per cent greater in the Smith diet than in that fed by Gies in Experiment 1, and somewhat greater in the Smith biscuit itself and 6 per cent less in the diet than in the Gies Experiments 2, 3 and 4. This 6 per cent less was of course much more than offset by the seven day feeding of Smith as against the six day feeding of Gies and also by the fact that the period of time over which the feeding extended was much greater in the Smith experiments. Hence, although from Gies' conclusion, the s.a.s. baking powder residue was possessed of a direct toxic action, sufficient to produce a chain of symptoms consisting of diarrhea; nutritional ineffectiveness whereby in general there was only 60 per cent of growth obtained, as compared with the growth of the animals fed other biscuit; marked interference with reproduction; impaired fecundity; and impaired nutrition of the offspring; on the contrary the effect produced in Smith's experiments, extending over longer periods of time and with greater quantity of the s.a.s. baking powder residue, was not any direct toxic action but a slightly lower rate of growth (Group 4) without any evidence of illness whatsoever. Indeed, the effect of Smith's diet was clearly a slight undernutrition, or a slight growth stunting without direct toxic action. Moreover, the very symptoms described by Gies as occurring in his s.a.s. baking powder groups were also the symptoms of undernutrition. They were not limited to growth stunting, however, as was Smith's effect, but represented a stage of nutritional ineffectiveness beyond that of growth stunting, effects which would occur when food deprivation reduced the resistance of the animal to a degree where infections would supervene. That Gies' animals had reached this stage of food deprivation is apparent from his observations that their growth was only 60 per cent. Stunting to this degree can be produced by limiting a particular quality of protein, but in such experiments there is no limitation of food ingestion in general.

Gies argues against deficiency of nutrients because the nutrients were practically the same in kind and proportion in all the biscuits, while only the s.a.s. baking powder rats exhibited the illnesses and effects described. This does seem to indicate that there was some special factor in the s.a.s. baking powder biscuit that was not present, or at least not to the same degree, in the other biscuit; but it is also apparent that it was not present with Smith's s.a.s. baking powder diets, although the latter contained a greater amount of baking powder residue.

There is one explanation that immediately suggests itself, i.e., that the tartrate biscuits yielded less growth than the standard because the excess of tartrate rendered such food less appetizing and in consequence the animals ate somewhat less of the tartrate than of the standard diet. The same was true of the sulphate diet. The s.a.s. baking powder residue in concentrated form is possessed of a decided taste, which is not detectable at all in ordinary household use, to be sure, but in the excessive amounts employed in the Gies and Smith experiments manifested itself in a very decided way. Hence, such food was eaten still less abundantly by the rats than was the tartrate and with the ingesta lessened beyond that of the tartrate, the growth was less. In the Smith experiments, this was true on the average and with 4 out of 5 animals, the fifth apparently finding the diet less objectionable as regards taste. This produced mere stunting of growth, to a limited degree, because the diet was not nutritionally deficient in kind.

In the Gies experiments, with the animals ingesting a diet inherently deficient in protein quality and quantity, in mineral ingredients and in vitamines, such further reduction of the diet from the taste of the excessive s.a.s. baking powder residue produced illness as well as growth stunting, not from any toxic action of the s.a.s. baking powder residue, as Gies believed he had demonstrated, but from the deficiency of nutrients consequent to the reduction of the already inherently deficient dietary. That it did not occur in the tartrate and phosphate animals was simply due to the fact that those foods revealed less to the taste the excessive residues which they contained and hence were eaten more abundantly.

It is fair to call attention here to the fact that the labels of tartrate and phosphate powders recommend the use of two teaspoonfuls of the tartrate powders, while labels of s.a.s. powders generally recommend the use of one teaspoonful. If these relative proportions of the different powders had been used in these experiments it is not unreasonable to expect that the comparative results would have shown a variation in the opposite direction from that actually observed.

It is clear that the facts did not at all justify the conclusions of Gies of "direct toxic action." Indeed the experiments herein reported definitely disprove any such action. That nutritional ineffectiveness could be and was in Gies' experiments promoted to such a degree through thus exaggerating nutritional deficiency may be admitted.

EXPERT OPINIONS

The following opinions on the foregoing investigations were given before the Federal Trade Commission in the hearings on Docket 540.[2]

Criticism by Steenbock. Harry Steenbock, professor of agricultural chemistry, Wisconsin State University, studied the testimony as well as one of the exhibits submitted by Dr. Gies, the testimony of Professor Heft and the testimony of Mr. Karshan covering the first four experiments of their work, the testimony of Dr. Hawk and of Clarence A. Smith, and the criticisms of Dr. Gies' work by Dr. E. E. Smith, in order to ascertain what factors might have contributed to the failure of the growth or the failure of reproduction, or the failure of the raising of the young. He did not agree with Dr. Gies' conclusions nor would he himself draw conclusions excepting that the evidence submitted was insufficient. The experiments were altogether too few in number. The use of a special ration one day a week was entirely indefensible. The lack of consumption of the diet by the "alum baking powder" rats might have been due to the lack of palatability rather than to toxicity. Muscular tremors, awkwardness of gait, and lack of coordination in Dr. Gies' animals did not mean anything to him as far as toxicity was concerned. A spasm or tremor could be a reaction of multiple origin. It was impossible to state whether the animal ate because it grew or grew because it ate. A large animal needs a greater intake of food than a smaller animal, so naturally as one has a larger animal, the consumption shows a greater food intake. Dr. Gies judged growth by increase in weight. Growth to the analytical mind meant an increase in length, in width, or in depth. No statement was made with respect to the appearance of the animals. As to activity, that was something which could be induced or prevented, dependent upon the manner of handling the animal. Partial failure of reproduction was not necessarily evidence of toxicity. A rat which does not grow normally never reproduces as readily as a rat which grows normally. These rats did not grow normally no matter what baking powder they received. This might have been due to various reasons; protein deficiency, or vitamin deficiency. Dr. Gies' work approached the problem from one point of view, using just one particular ration. Prof. Steenbock could not in any manner justify for himself a procedure where 20 rats fed on "alum baking powder" should be given the responsibility of deciding whether such powders were toxic. It would be necessary to approach such a problem from many angles, and if all the evidence should then converge to point to toxicity as a fact, the conclusions would be justified.

There was nothing new introduced by Dr. Hawk's experiments excepting that he made conditions even more unsatisfactory by introducing the practice of giving the rats meat on Wednesdays. The prior criticisms applied equally to the work of Dr. Hawk and his assistants.

Dr. E. E. Smith's report brought out this matter of palatability as entering in determining the outcome of Dr. Gies' experiments. He used a ration satisfactory for allowing growth to take place normally. Dr. Steenbock did not know whether the evidence which Dr. Smith presented was sufficient to demonstrate the non-toxicity of the baking powder. It was all right as far as it went but was not sufficiently extensive. Just as Dr. Gies used a ration too deficient so Dr. Smith might have erred in using a ration that was too satisfactory. It seemed absurd to place the responsibility on such a few experiments. Hundreds of rats are used to determine some theoretical point; and here in a matter of tremendous economic bearing, of physiological well-being of many people, it seemed that thousands of animals should be used, instead of a few paltry dozen. It was a very practical thing to determine the palatability of the food.

"Most of the statements that I have made have back of them extensive experiments and extensive experience gained from experiments of an allied nature."

Criticism by Mendel. Lafayette B. Mendel, Professor of Physiological Chemistry in Yale University, had employed rats as experimental animals since about 1910 and recalled that he had records of at least 12,000 animals in studies on nutrition. He had read the testimony of Dr. Gies, Professor Heft, Mr. Karshan, Dr. Hawk and Clarence A. Smith and Dr. E. E. Smith, and casually that of Dr. Hammett, also Exhibits Nos. 254, 327 and 328, which are, respectively, the work of E. E. Smith, Hawk and collaborators, and Gies and collaborators; and noted that Dr. Gies' fundamental thesis in connection with the problem was "whether aluminum given in aluminized food over a long period of time would have any effect on the growth of animals and the fecundity of females."

He found several items of evidence as follows:

1. A less rapid growth of rats on a diet composed of biscuit prepared with so-called "alum" baking powder than with similar biscuits prepared with tartrate and other powders.

2. Some of the rats fed on the so-called "alum" baking powder biscuit showed certain symptoms of abnormality, namely, diarrhea, incoordination, tremors, spasms, paralysis, and the like.

3. Indications of so-called lower fecundity.

None of these claims were properly warranted by the type of evidence that he had discovered in the record and specifically referred to. It was unwarranted to infer that aluminum in the food was the cause of a less rapid growth unless it was demonstrated that the animals actually ate as much food as those that were on the other diets. In every series excepting one it was clearly shown that the animals on the so-called "alum" food did not eat as much as the animals on the other foods. It was unwarranted to conclude that the biscuits prepared with the so-called "alum" powder had no nutritive value unless they could be compared on a precise, equivalent intake basis. To illustrate, suppose 2 individuals were served with milk as the major part of their food, the one cow's milk, the other goat's milk, the nutritive values of which were essentially the same. The person might drink all of the cow's milk offered him, say 2 qt. per day, and the other might dislike the peculiar odor or flavor of the goat's milk, drink very little of it, and therefore, fail to make the gain on the goat's milk that his counter-subject did on the cow's milk; or he might even lose weight. It would be unwarranted to argue that the goat's milk contained a poison because he failed to gain on it. It could only be used as evidence of some toxic substance, if both persons had taken the same amount of nutriment from the two sources. The failure to make expected gains was not necessarily due to a toxic ingredient of a diet but might be due to insufficient intake of one or more dietary essentials. For example, in the case of the person above taking insufficient goat's milk, the lack of lime, rather than necessarily any toxic factor in the goat's milk, might be a decisive factor producing certain results. There was no warrant for the unequivocal conclusion that the toxic substance in the case of these rats was aluminum or sulphate or both. Professor Mendel also concluded that the results of these experiments gave no warrant to conclude that the particular products used containing aluminum were inferior from the hygienic standpoint. As to symptoms of abnormal ty, diarrhea was not a regular accompaniment of the "alum" food, and the data given failed to show facts essential to give a clue as to whether diarrhea was associated with a poison or with some other factor. The symptoms of incoordination, such as the falling over of the animals or the inability to use their legs properly, are not at all unusual in rats that failed to get proper nourishment. Lowered fecundity is not necessarily an indication of dietary toxicity. Similar results attend deficient food intake. If the "alum" rats ate less food, it was to be expected that they might show any or all of the phenomena noticed by Dr. Gies or Dr. Heft. The implication that marked toxicity in the mothers was due to the ingestion of "alum" is not warranted. Was it a diet containing a poison or a failure to eat well that was the cause of the symptoms? Where the growth is less rapid, the intake is always less. It happens that in the first Gies series, during the first month they ate equally well and they grew equally well. When the dose of the powders was increased 50 per cent they still grew equally well until the milk was withdrawn, but then they ate less and grew less. It is quite as justifiable to say that they did not grow as well because they did not eat as well as to say that they did not grow as well and therefore did not eat as well. It is quite as proper to assume that, with the milk withdrawn, the flavor or palatability of one lot of biscuits was sufficient cause of the decreased food intake, as to assume that there was poison there so that they could not eat. In the presence of an abundance of milk this flavor factor may not have shown up, while in the absence of milk and with an exclusive diet of biscuit it might do so. In the third series of Hawk's experiments the mortality of the "alum babies," if he might call them so, was less than the mortality of the "tartrate babies." This was not brought out in Hawk's conclusion, but Professor Mendel could evaluate that kind of thing to show that tartrate produced less valuable stock than "alum" did. The failure to eat well in the case of rats fed with "alum" may have been due to the relative unpalatability of the biscuits, owing to the amount of leavening agent used. The failure to eat *per se*, in his judgment, was not a proof of the conclusion of Dr. Gies. There were two variables, the variable composition with reference to the leavening agent, and the variable food intake. Each of these factors must be evaluated in arriving at the conclusions. A type of experiments could have been formulated which would have entirely eliminated the inequalities of food intake.

A competent observer cannot necessarily determine whether a substance is palatable by taking note of the avidity with which a rat or rats seeks the particular substance. Mendel knew of instances where rats were influenced by relative palatability of foods, where he himself probably would not have discovered the difference. The rats made selection between two foods that to his taste were the same. It is not an uncommon phenomenon for an animal to starve itself against a food which it does not like. It is no more puzzling to account for that than that a dog could find its way across a field by following a track. Those who are experienced in this field of work would recognize at once the undesirability of so constituting a diet that a minimum of any one of the factors essential to the nutritive well-being might simulate in the animal an effect of a harmful result from an additional factor that might be incorporated in the diet. It would be an ill-conceived experiment in which the diet was so nutritionally poor that if an added substance were introduced for investigation, one could not tell whether it was inherent poorness of the diet or the substance added that produced the deleterious effect. Bread for a rat is an inadequate diet. There is a difference between an optimum diet and a diet which contains all the essentials. On a given optimum diet rats grow better, breed sooner and produce more young and more thrifty young than on the same diet in a different proportion, which is an adequate diet. The best growth of the rats of any series, he was here considering, was not up to par. The diets were inadequate in every case. The best rats would have undoubtedly grown faster if they had been on a better diet.

FURTHER STUDIES

Amy L. Daniels and Mary K. Hutton have recently[3] published studies extending over a number of years on the mineral deficiencies of milk as shown by growth and fertility of white rats. The authors had observed that "rats fed exclusively on milk, either raw or quickly boiled, seldom reproduced" and "where reproduction did take place, the young in general were born late and only a very small percentage survived the suckling period. A third generation on milk alone were never raised." This suggested that milk was deficient in substances essential to normal physiological processes.

Similar results were obtained with whole milk powder. Supplementing milk with such substances as cereal diluents, wheat embryo extract, lactose, beef extract, cornstarch, cod-liver oil, and, later, cystine, and various calcium salts, failed to improve the nutritive value of milk. With the addition of iron citrate, the number of fertile animals was somewhat higher, yet considerably below normal, and the third generation never reproduced. With the further addition of iodine slightly better results were obtained, with one group a fourth generation being produced which grew normally for three months and then only very slowly and failed to reproduce further. Considered as a whole, there was no uniformity, either in fertility or in the number and viability of the young.

The report of Ruhräh,[4] lauding the value of soy bean milk mixtures in cases of undernutrition in infants, suggested its trial. The addition of 7 gm. to 10 gm. of soy bean powder, cooked, and added to 1 liter of boiled milk, produced surprising results. Young in the first generation of rats were obtained in thirteen weeks; second and third litters followed in quick succession; and these in turn bore young which were successfully raised. Animals'were then fed milk to which the ash of soy beans had been added. All females in the first generation were fertile, the first litters being born before the age of six months (which time was the standard adopted for fertility measurement and number of young born), were normal in number and the individuals of average size. Moreover, 76 per cent of these young were successfully suckled. With milk alone in the first generation only 12.5 per cent of the females were fertile up to six months, few young were born and none of these lived. In the second generation on milk alone, only 1 of 5 females was fertile up to six months and none of the 5 young born in this group was raised. In contrast to this, the 10 females of the second generation on the soy bean ash milk mixture produced 67 young, all of which were raised. In the three successive generations on milk and soy bean ash, an average of 6.3 young per female up to six months was successfully suckled.

An analysis of the soy bean ash showed that aluminum and silicon were present. Osborne and Mendel[5] had reported significant traces of aluminum, fluorine and manganese in milk and had incorporated these in their purified rations in the proportions found. It seemed possible that some or all of these might be essential to physiological processes and, although they were present in cow's milk, might be in too small concentration for reproduction and rearing of the young of the particular species of animal under investigation. Since silicon was found in all vegetable and in some animal tissues, it was thought that this also might be essential to physiological processes.

Accordingly, a thick starch paste, containing a mixture of aluminum potassium sulphate ($AlK.(SO_4)_2.12H_2O$), sodium fluoride (NaF), sodium silicate (Na_2SiO_3) and manganese sulphate ($MnSO_4.4H_2O$) was added to milk in such a proportion that each animal received daily 1.5 mg. of each salt, a purely arbitrary amount. One drop of a 2 per cent solution of sodium iodide and 3 drops of a saturated solution or iron citrate was further added to each 200 cc. of the milk.

On milk thus supplemented with these 4 mineral constituents, the rats developed admirably; young were born between the fourteenth and sixteenth weeks, were vigorous and developed normally. In each succeeding generation, all females were fertile, first litters being produced early and a large proportion of the young was raised, an average of 80 per cent in the 5 generations. Six generations were produced with every indication of continued growth and reproduction. Two females of the sixth generation bore young at three and one-half and three and three-fourth months, respectively. From these results, it seemed that the substances lacking in milk were included in the above mixture of four salts.

That all substances necessary for normal growth and reproduction were in milk, but in too low concentration, seemed evident, not only from the work of Daniels and Hutton, but from that of Anderegg[6] and of Mattill, Carman and Clayton.[7] In both of these investigations, animals fed milk low in fat did not exhibit marked reproductive failure. Apparently, when more milk was taken, as would be the case in rats fed a low fat milk, enough of the various essential substance was obtained for the production of a next generation.

Trials were then made by Daniels and Hutton with the separate salts above named with combinations of two or more, iron and iodine, as well, being constantly added to the whole milk. In evaluating the results, the high mortality of the young in the successive generations of many groups, even when fertility was high, led to the conclusion that the average number per female of young raised within a six months' period was a better criterion of the potency of the added salts than was the fertility. Judged by this standard, the animals receiving the mixture of the four salts outclassed all the other groups.

The addition of aluminum $(AlK(SO_4)_2.12H_2O)$ in one series and of sodium silicate (Na_2SiO_3) in another, resulted in a higher percentage of fertile animals and a fair number of viable young through three generations. When these two salts were used together, all females in the first and third generations were fertile, whereas those in the second and fifth reproduced very late. With the exception of the first, the number of the viable young in each generation was very low. The addition of either manganese or fluorine to the aluminum and silicate mixture resulted in a high percentage of fertility and a fair number of viable young. Whether manganese or fluorine was essential was not clear. Alone in the concentrations used, they appeared to be toxic. Combinations of aluminum and manganese and of aluminum and fluorine were also unsuccessful. In neither case were there young in the first generation. A combination of aluminum, manganese and fluorine, however, carried on to the fifth generation. With the exception of the fourth generation, fertility in those was slightly below normal, and the number of viable young was considerably below the average in each generation.

Aluminum and sodium silicate appeared to be necessary. These, however, must both have been so incorporated in the diets that they were both available. The happy results produced where manganese and fluorine were added to the aluminum and silicate mixture might have been due to the fact that the aluminum and silicate were thus made more readily available; on the other hand, traces of manganese and fluorine might have been necessary for physiological processes.

As bearing upon the results of Daniels and Hutton, recent experimental observations by Helen S. Mitchell and Lola Schmidt[8] in the Nutritional Laboratory, Battle Creek Sanitarium and College, Battle Creek, on "The Relation of Iron from Various Sources to Nutritional Anemia" are significant. In providing anemic rats for their studies, they were presented with the difficulty that reproduction on milk diets was poor and mortality of the young high. They report that when milk was used as the basic diet, the addition of traces of manganese, fluorine, silicon, aluminum and iodine, in the quantities suggested by Daniels and Hutton, seemed to promote better reproduction without affecting the anemia in the young rats (iron was not fed). This observation is in harmony with the results of Daniels and Hutton and to a degree confirmatory of their conclusions.

CHAPTER XV

DISCUSSION

In the preceding chapters facts have been presented showing the action of a variety of aluminum compounds when brought into contact in various ways with a wide range of living tissues. These include the tissues of plants and animals, the latter ranging from lower forms of animal life to the higher forms, including man. It is now our problem to see how these facts apply to the subject at hand, namely, the influence of aluminum compounds, as they occur in food, upon the health of man.

Unfortunately, this question has become controversial by reason of conflicting commercial interests. In order to arrive at correct decisions, it is important to divert the mind from such extraneous influences and direct the attention solely to an impartial weighing of the evidence bearing upon the hygienic questions presented. Only in this way are trustworthy conclusions to be reached and the interests of the consuming public served. It is not fair to the problem involved to approach it blindly holding to preformed opinions and unwilling properly to evaluate new evidence. At the same time open-mindedness is not to be construed as an indication of vacillation but rather of a willingness to give proper weight to any new evidence that at any time may be presented.

The trend of the arguments of those who hold the opinion that aluminum salts in food are injurious is that under certain conditions aluminum compounds may produce injury in the body. Those who have reached the opinion that aluminum salts as they commonly occur in present-day foods do not render such foods unwholesome cannot and do not question such a possibility, but they point out that the possibility exists only under conditions widely different from those presented by the actual use of aluminum compounds in food.

They base their opinion upon what they believe to be the fact that the conditions with injury do not pertain with the foods in question and that the latter are separated from the former by a substantial and wide margin of safety. This is not a novel position but one that is taken relative to other food ingredients—it is, in fact, general, and not the exception. The body is an intricately adjusted mechanism. Any substance, no matter what it is, indiscriminately thrown into the body machine without regard to quantity or concentration of the substance or the part of the machine into which it is thrown presents the possibility of producing injury or perversion of function.

It happens that at the time of this writing, the press contains an account of a case of attempted suicide by drinking tincture of iodine, a deadly poison. There is not the slightest evidence that a few drops of the tincture, well diluted, would produce in any degree the effect produced by the large amounts of the concentrated tincture as taken. Even potassium iodide, in concentrated solution, produces gastric irritation, yet in limited quantities and when sufficiently diluted with water or with certain food admixtures is entirely without deleterious effect. In fact, a certain small amount of iodine is a normal and essential food ingredient without which the body cannot properly grow or function.

Common table salt is another illustration often referred to in this connection. Under some circumstances and in certain amounts it may be lethal; while under other circumstances and in smaller amounts it is essential to the body.

An example closely analogous to the relation of aluminum compounds to the body is found in the compounds of iron. Experimental observations of the action of the compounds of iron, so far as they have been conducted in parallel, show in general the same effects as have been observed in the action of compounds of aluminum. Under the unnatural conditions of such observations iron has quite as inimical an action as has aluminum. There is reason to expect this, as the two elements are very closely related in their chemical behavior. Iron is an essential constituent of the body. There are some reasons for believing that aluminum is likewise an essential constituent of the body. Iron, when administered in massive amounts or when introduced into the body through unnatural channels, may have an injurious action, but this does not indicate that iron in food is essentially injurious. It produces injurious effects when misplaced in the machine, so to speak, or because of an abuse of the quantitative relation. The fact that when misplaced or when excessive in amount, it acts unfavorably, merely indicates the objection to misplacement or excess of quantity; it is not an evidence of harmfulness of amounts in food that are well within the limits to which the body adapts itself without injury.

The real question is whether a particular amount serves a useful purpose and whether the amount is well within the limits of safe consumption. The question is not whether a food ingredient may by abuse, either in the amount consumed or because of an unnatural channel of administration, produce untoward effects, but whether under the conditions of actual consumption it does in fact do so. That should and does constitute the object of inquiry in the discussion that follows.

Disregarding views formerly presented that have been abandoned, the claims now put forward in support of the view that added aluminum compounds in general, and aluminum compounds derived from baking powders containing s.a.s. in particular, render food unwholesome or injurious to health, may be summarized as follows:

Claims of Unwholesomeness. 1. The claim that aluminum may combine with various essential constituents of the food, viz., phosphates, food accessory substances or vitamines, and thereby may deprive the body of them "just as if they had not been present in the food at all."

2. The claim that in the gastrointestinal tract aluminum compounds may act as irritants and impair digestive efficiency.

3. The claim that aluminum is absorbed into the lymph and blood and accumulates in certain tissues of the body, especially the liver, thyroid and adrenal glands.

4. The claim that aluminum is a protoplasm poison.

5. The claim that absorbed aluminum has a deleterious effect upon the blood corpuscles and produces injury, notably

to the kidneys and nerve tissues, with a tendency to the production of anemia and a lowered resistance to infections and disease.

6. The claim that it retards growth and diminishes fecundity and fertility.

I. THE CLAIM OF COMBINATION WITH ESSENTIAL FOOD CONSTITUENTS

That the aluminum ion combines with the phosphoric ion to form insoluble or difficultly soluble phosphates is well known. Further, it has been demonstrated that the ingestion of aluminum, at least in certain forms of combination and in sufficient quantity, tends to deflect phosphate elimination from the urine to the feces. The question presented thereby is whether this constitutes an injury to the body and if so, whether it is effected by depriving the body of phosphate.

Deflecting Phosphate. In the first place, it is well to call to mind that this is not a property peculiar to compounds of aluminum but one shared as well by calcium and probably by iron and other similar elements. Thus, if because of this action aluminum inflicts injury to the body, then in precisely the same way calcium might inflict an injury. However, it does not appear that with calcium this constitutes a very ominous danger; for calcium in its various forms is ingested *ad libitum,* both as a natural and as an added constituent of food, not only without apparent injury but without even a sufficient possibility of injury so that attention is commonly called to the matter.

In the second place, is it possible that a certain amount of deflection of phosphate from the urine to the feces, an amount normal for calcium, is without injury but that harm arises from the additional phosphate deflection imposed by added aluminum in food?

This raises the question to what extent added aluminum in food deflects phosphate. This was made the subject of direct and exact observations by each of the investigators of the Referee Board. In considering their results, it is well to have in mind that they regard a 75 mg. per day as the maximum amount of aluminum ingestion from the occasional use of s.a.s. baking powder food for the days of consumption of such food; (b) 150 to 200 mg. per day as a large amount that would be ingested only under very unusual conditions, where all the flour was leavened with a straight s.a.s. baking powder; and (c) an amount in excess of this such as would be taken only when a person subsisted on such food in very large amounts and to the exclusion of other foods.

How much do these amounts deflect phosphate elimination from the urine to the feces? Chittenden studied this problem (p. 97) by determining the nitrogen-phosphorus ratio in the urine of 8 subjects each of whom ingested, over a period of one hundred and thirty days on a fixed diet, amounts of baking powder bread containing in the first ninety-day period 68 mg. of aluminum daily, and in subsequent ten-day periods respectively 140 and 186 mg. daily, and in a final thirty-day period 258 mg. daily. There was no change in the nitrogen-phosphorus ratio, from which fact the conclusion was reached that there was no deflection of phosphate from the urine to the feces.

Long studied (p. 105) the elimination of phosphates in the urine of 6 subjects on a controlled diet to whom was administered for seventy days amounts of straight s.a.s. baking powder residue, containing in the first thirty days 224 mg. daily, and in a subsequent 10-day period 447 mg. daily of aluminum; and in three ten-day periods, respectively, 447, 670 and 1118 mg. daily of aluminum in the form of the baking powder residue from which the sulphate had been removed. With the 224 mg. of aluminum administration, there was little or no reduction of the urinary phosphate, as indicated by the nitrogen-phosphorus ratio, in this respect his results agreeing substantially with the results obtained by Chittenden; but with the larger amounts there was definite reduction up to one-third of the phosphate present in control periods on a similar diet. The urinary acidity, dependent in a considerable measure on the phosphates, was reduced somewhat in all the aluminum periods, the maximum reduction approximating the maximum phosphate reduction.

In a subsequent series less accurately controlled as to diet, Long (p. 107) again observed reduction in urinary phosphates, though in a somewhat lesser amount.

Taylor (p. 119) conducted exact studies of the phosphorus intake and outgo in 8 subjects, 7 of whom took potash alum (P.A.S.) or S.A.S. An amount containing 100 mg. of aluminum (P.A.S.) daily in 6 subjects showed no deflection of phosphorus from urine to feces in 4 and 9 per cent in 2 subjects. Increasing the aluminum ingestion (S.A.S.) up to double the amount showed phosphate deflection in the 6 subjects, the maximum decrease in the urine being 12 per cent. A daily amount of 0.298 mg. in 7 subjects only slightly increased deflection, if at all, the figure rising to 20 per cent only in one instance. In another series of observations, when the aluminum was administered as hydroxide or as baking powder residue, the amount of phosphate deflected was less than when S.A.S. was administered, even though in some instances the amount of aluminum ingested reached double the amount in the S.A.S. series. In no instance in either series did the ingestion of the aluminum compounds increase total phosphorus elimination, an exact or usually a slight positive phosphorus balance, as judged by the daily average of the series, being maintained.

Of even greater interest and importance were the observations of Taylor (p. 127) on 2 subjects whose diets were extremely low in phosphorus, 0.37 and 0.07 gm., respectively, in place of 1.5 gm. daily. In these instances there was neither deflection of urinary phosphate by aluminum hydroxide administration (660 and 540 mg. Al, respectively) nor any increase of total phosphate elimination. That is to say, under this condition of a high negative phosphorus balance, a condition under which the matter of disturbances of metabolism and elimination might be of some consequence to the body, there was no disturbance of phosphorus elimination and no deflection whatsoever from the urinary channel of elimination, even though very large amounts of aluminum were administered.

It was upon this evidence that members of the Referee Board reached the conclusion that the aluminum of the diet did not in any discernible way disturb the normal flow of the total phosphorus metabolism and that any tendency of aluminum to deflect phosphorus elimination from the urine to the feces was not objectionable. Comparing the figures of the two observers, it will be noted that Taylor obtained an inconsiderable amount of deflection while with the same amount of aluminum ingestion Chittenden obtained none. It is possible but not probable that this difference was a mere coincidence. The explanation may exist in the different form in which the aluminum was administered by these two investigators. Chittenden gave his in the form of s.a.s. baking powder food while Long and Taylor administered their aluminum independently of the food, simply as s.a.s., as the hydroxide or as s.a.s. residue. It is highly probable that the combination of the aluminum that has been baked in the food is quite different from the hydroxide or s.a.s. residue prepared independently of the food. Other considerations, namely, the fact that at most very little and probably none of the aluminum is absorbed and that in the absence of phosphates in any considerable amount in the alimentary tract there is no diminution of urinary phosphates, indicate that there is no true deflection of phosphates so far as the endogenous phosphates, derived from metabolism, are concerned, but that what really occurs is a fixation in the alimentary tract of a certain amount of exogenous phosphates present in the food, presumably as inorganic phosphate, so that they leave the body as aluminum phosphate in the feces without having been absorbed. If this be true, it offers an adequate explanation of the apparent deflection in Taylor's observations where the aluminum was simply in the form of s.a.s. or the hydroxide, while there was no diminution of urinary phosphate and hence no deflection in the subjects under Chittenden's observation where the aluminum was baked in the food.

Urinary Acidity. Since the amount of phosphate in the urine ordinarily determines to a considerable extent the degree of urinary acidity, it is to be expected that the diminution of phosphates would be accompanied by diminution of the acidity. Chittenden does not report the urinary acidity but the observations of Long are very definite upon this point. In the first series (p. 106), and regarding the mean acidity for the twenty-day control periods of all subjects as 100, during the administration of the residue of 2 gm. S.A.S. (0.2236 gm. Al) there was a mean diminution of acidity of 23.5 per cent; of 4 gm. (0.4472 gm. Al), 25.5 per cent; of 6 gm. (0.6708 gm. Al), 22 per cent; and of 10 gm. (1.118 gm. Al), 18 per cent. In the second series, observations on the acidity (p. 107) included, in addition to the 4 subjects to whom was administered S.A.S. (a), parallel studies on the same number of different subjects to whom was administered calcium acid phosphate residue (b), and Rochelle salts (c), with the following results (figures in parenthesis show per cent decrease acidity):

TABLE CXV

Period	Amount of S.A.S. residue or phosphate or Rochelle salt equivalent	Acidity of urine (means)		
		S.A.S. subjects	Calcium acid phosphate residue subjects	Rochelle salts subjects
1	None	2.21	1.75	2.05
2 to 4	None	....	*1.55(11)	
	2 gm.	1.91 (14)	....	1.32 (36)
5 to 6	None	....	*1.55 (11)	
	3 gm.	1.88 (15)	....	1.03 (50)
7 to 8	† 2 gm.	....	1.48 (15)	
	4 gm.	1.56 (29)	....	0.53 (74)
9	None	1.75 (21)	1.70 (3)	1.73 (18)

* Physiological salt solution administered. No calcium acid phosphate.

† Calcium acid phosphate equivalent to 2 gm. S.A.S. administered.

Keeping in mind that even the residue from 2 gm. S.A.S. is triple the ordinary daily consumption from the use of straight S.A.S. baking powder, double the large consumption and equal to the consumption in extraordinary amounts, it must be concluded that by the consumption of straight S.A.S. baking powder food:

1. Urinary acidity is not markedly reduced.

2. Any change is within the limits of physiological variations.

3. It produces only about one-third of the diminution of urinary acidity that is produced by an equivalent amount of Rochelle salts.

Discussion. There is another interesting suggestion that comes out of the foregoing experimental studies of phosphate deflection and diminished urinary acidity, to which attention need only briefly be given.

It has doubtless occurred to the reader that the foregoing data presented do not indicate deflection of phosphate or reduction of urinary acidity proportional to the amount of S.A.S. residue or aluminum hydroxide administered. This raises the question whether all kinds of exogenous phosphate become fixed in the alimentary tract by aluminum hydroxide. In addition to inorganic phosphates, there is present in food phosphorus-containing proteins, phosphatids (phosphorized fats) and phosphoric acid esters of carbohydrates and related substances. It may well be that none of these enters into combination with the aluminum, or does so only to a limited extent, leaving only the inorganic phosphates to enter into combination and thereby escape absorption and give the appearance of being deflected in elimination.

Further, it may well be that, in the foregoing experimental studies the major part of the inorganic phosphates combined with the lesser quantities of S.A.S. residue and the ingestion of such residue in greater amount could not proportionately increase apparent deflection because of the lack of phosphate remaining available for fixation. It would require additional data to prove or disprove this suggestion, nor is the matter of importance to the subject at hand.

From the foregoing it is apparent that in the amounts in which aluminum might occur in the diet from alum in pickles, from S.A.S. in baking powder and from other sources phosphates are not withdrawn from the body; phosphates are deflected from the urinary to the fecal channel of elimination to only a slight extent, if at all, the investigations with the actual occurrence of aluminum in food showing no deflection; and urinary acidity is not lowered in any substantial amount, since the differences are within the limits of ordinary physiological variations, and only a third as much as would result if tartrates were substituted for s.a.s. in equivalent amount.

Notwithstanding the facts as presented, Prof. H. Gideon Wells in recent testimony[1] referred to Taylor's work as showing that "aluminum withdraws phosphates from the body." Reference to Taylor's figures (pp. 119, 125) shows that while there was apparent deflection of phosphate from urine to feces, there was no withdrawal of phosphates from the body, a balance being maintained throughout in spite of the administration of large amounts of aluminum compounds; and that when, by the selection of a diet very low in phosphate (p. 127), a large negative balance had been established, such negative balance was not increased by the administration of large amounts of aluminum compounds, nor under these circumstances was there even a slight deflection of phosphates. Further, Professor Wells refers to the work of Leary and Sheib (p. 248) as "confirmatory of the fact that administration of aluminum deflects phosphate from urine to feces with interference to growth," but omits to mention the qualification by Leary and Sheib with reference to the deflection of phosphate: that it "of course has no special bearing on the question of the behavior of aluminum compounds as used in the diet of man, as the amounts of aluminum used were excessive." In fact, they used a minimum daily amount of about 100 mg. Al per kilogram body weight in the case of puppies and 500 mg. to 3000 mg. daily with rats. Such amounts are from 100 to 2000 times the equivalent amounts that would occur in the diet of man, and, as the authors carefully state, are inapplicable. Where testimony has to do with the question of aluminum compounds, as used in the diet of man, the citation of Leary and Sheib's general conclusions without their qualification of the inapplicability of the results to the diet of man is, therefore, unfortunate and tends to be misleading.

The subject of phosphate elimination following the ingestion of aluminum and particularly s.a.s. should not be dismissed without calling attention to the fact that this product is used in baking powders at the present day almost if not wholly in conjunction with calcium acid phosphate and that the more widely used brands contain more phosphate than is required to combine with all of the aluminum present. Hence, even the deflection noted with straight s.a.s. powders, which probably rarely if ever occurs in the actual consumption of food prepared with such powders, and certainly never to a degree that is to be regarded as objectionable, gives no reason to justify the conclusion that similar deflection occurs from food prepared with s.a.s. phosphate powders. Under these conditions Long (p. 111) found no diminution of urinary phosphates.

Combination with Food Accessory Substances. The suggestion has been made that aluminum in food may combine with food accessory substances or vitamines, this idea being expressed thus: "Probably a more serious result will be that it will unite with various essential constituents of the food, present in small quantities, substances called food accessory substances," etc. In view of the fact that we have been living during the last few years in an era of vitamines, so to speak, it is natural, quite proper, and even important to consider whether any such combination does in fact occur. A belief in such an occurrence is not only merely speculative but opposed to the evidence afforded by the most gigantic test of which it is possible to conceive, namely, the ingestion of food prepared with s.a.s. baking powders over a period of decades by millions of people. Notwithstanding this test, vitamine deficiency is an exceptional occurrence and when it does occur is in the great majority if not in all instances explained by the ingestion of a diet originally deficient in vitamines. The fixation of vitamines by the combination of substances in the food has, so far as the writer knows, not only never been proved but is not seriously considered in this book or elsewhere, as a possible occurrence. The vast experience cited is wholly opposed to such a view and practically disproves it.

2. THE CLAIM OF GASTROINTESTINAL IRRITATION

The suggestion that in the gastrointestinal tract aluminum compounds may act as irritants and impair digestive efficiency is based upon the following considerations:

(1) The effect of alum or other forms of aluminum administered without food or without having been cooked in food.

(2) The effect of massive doses of baking powder residues in food.

(3) The inhibiting effects of baking powder food upon enzyme activity.

(4) An alleged toxemia from increased formation or absorption of products of gastrointestinal fermentation and putrefaction.

That soluble saline matter in concentrated solution, whether salts of iron, salts of aluminum, salts of calcium, salts of magnesium, salts of potassium or salts of sodium, whether sulphates or chlorides, including common table salt, is capable of exerting an irritating effect upon the mucous surfaces of the gastrointestinal tract need not be questioned. This fact was recognized and clearly set forth by writers more than half a century ago. It is for this reason that a concentrated solution of any saline tends to induce vomiting. It is a property of the hypertonicity of the solution rather than the amount ingested; or, as Tardieu so well expressed it in 1875, "the concentration is of more importance than the dose."

The problem of the action of hypertonic solutions on the mucous surfaces of the intestinal tract is not presented by the conditions that arise by reason of the presence of aluminum compounds in food. It obviously cannot be maintained that hypertonic concentration does or can exist by reason of such presence; hence it does not follow that a small quantity acting for a long time may produce results at all comparable to those that might be produced by a large quantity acting for a shorter time, this latter view overlooking or ignoring entirely the factor of hypertonic concentration in the production of symptoms of irritation. All references of acute manifestations by reason of the

ingestion of alum or similarly soluble salts of aluminum dealing *only with hypertonic solutions* of these substances call for
no consideration whatsoever in connection with the question
under discussion. To maintain such a possibility is merely
in effect to confuse the issue and is wholly unwarranted.

It is of concern whether aluminum compounds in food do
exert a specific irritating action not dependent upon hypertonic concentration. Mallet (p. 51) was one of the early
investigators to attempt to determine this question. His
ingestion of massive doses of the hydroxide or of the phosphate, on the supposition that these were baking powder
residues, ofttimes on an empty stomach and with relatively
little water, with the production of oppressive sensations,
did not answer the question. If there had been admixture of
such residues with a relatively very large amount of the
constituents of dough and subsequent heating in intimate
contact with the same, as occurs in baking powder food,
not only would the dough have acted as a diluent but, it
would seem, must have changed the physical condition and
even chemical combinations of the baking powder residue.
In the presence of such changes it is not true, as recently
claimed, that because milk acts as an antidote its administration in animal experiments with baking powder food
would vitiate experimental results. Milk is an antidote
because it is a convenient and soluble form of protein that
may be readily brought into intimate contact with a metallic
salt in the stomach, and may enter into combination with it,
thus preventing or diminishing its ionization and hence its
irritating action on the walls of the stomach. It may be that
in experiments like Mallet's, milk or other form of soluble
protein by so acting would have prevented the oppressive
sensations which he reported. However, as will be seen far-
ther on, under the physiological conditions that exist with
the ingestion of food, the small quantities of aluminum in
the baking powder residue are insufficient to give rise to
concentrations of aluminum salts in solution sufficient to
have any irritating action. Moreover, not only have they
already been brought into intimate contact with the food
protein but also actually cooked with it. Thereby the oppor-

tunity for interaction of aluminum and protein has been afforded in the preparation of the food before it enters the stomach and the presence or absence of milk is not an important factor, contact with protein having been provided by the ordinary procedure of preparing the food.

Any claims of irritation of the gastric or intestinal mucosa by baking powder residues, based upon experimental observations made with aluminum salts other than those in which such compounds are as thoroughly mixed and cooked with food as occurs in the habitual preparations and ingestion of such food in everyday life, are clearly wholly unwarranted. Such claims are even more unwarranted than would be, for example, claims of an irritant action of many of the condiments, if their effects in food were to be predicated upon their direct action when fed alone instead of when intimately mixed with food in such dosage as occurs in their ordinary use. With baking powder residue there is not only the intimate admixture with food but the probable interaction with food ingredients, as described.

The Claim of Impairing Digestive Efficiency. There have been few experimetal studies of the direct effects upon the gastric and intestinal mucosa of s.a.s. baking powder residue in food. The writer's early observations (p. 62) of the comparative effects on gastric secretion of food prepared with and without such baking powders, showed no difference in the composition of the gastric contents following breakfasts of the respective foods. His comparisons of the test-breakfast gastric contents at the beginning and end of nearly five-month periods, during which baking powder residue in food was ingested in unusually large quantity, showed what appeared to be an appreciable increase in acidity in 2 experimental subjects. However, as has been pointed out (p. 208) in the case of the one subject upon whom previous observations had been made, the apparent increase was not appreciably greater than the observed acidity some two years before the experimental ingestion of such food. To attribute such variation to any irritant action of the food of the experimental studies is therefore wholly unwarranted. In view of this fact, it would seem necessary that a large series of such observations should be made, on subjects whose range of variation of the acidity of the gastric contents over a considerable period of time had previously been established in order to afford any competent direct evidence whether gastric acidity is at all influenced by the baking powder residue. Such observations have not been made but other considerations point against the conclusion of any such harmful effect.

In discussing gastrointestinal irritation, it is reasonable to take into consideration the concentration of solution that may come into contact with the gastric and intestinal mucosa. This is best determined by direct observations on the contents of the gastrointestinal tract, it being clear that evidence from solubility tests in vitro must give way to the concentration found actually to occur in the body. The following data bear upon this point.

TABLE CXVI

GASTRIC CONTENTS

Observers	Type of s.a.s. B.P.	B.P. grams per 100 gms. flour	No. of observations	Soluble Al, per cent as $AlCl_3$		
				Max.	Min.	Average
E. E. Smith..........	Straight	4.0	2	.0044	.0044	.0044
Mallet (p. 194).......	Straight	2.9	3	.035	.017	.028
Hawk and Smith (p. 195)..............	Straight	9.1	2	.036	.025	.031
Myers and Killian (p. 197)..............	Straight	1.42	5	.035	.0029	.018
	Straight	1.80	1	....	....	.0089
	Straight	1.90	1	....	....	.026
	Straight	1.20	1	....	....	.0069
(p. 196)............	Combination*	2.73	4	.022	.0033	.0098
(p. 196)............	Combination†	6.6‡	2	.040	.016	.028
	Combination†	1.6	4	.0073	.00057	.0044
INTESTINAL CONTENTS						
Myers and Killian (p. 198)	Straight	1.77	3	.00509	.00067	.00328

* Containing 4.19 per cent Al_2O_3.
† Containing 2.85 per cent Al_2O_3.
‡ 4 times the amount indicated on the label.

Assuming the soluble aluminum to be aluminum chloride, it thus appears that the maximum concentration observed in the gastric contents was 0.04 per cent, and that the usual concentration was materially less, while in the intestinal contents the dilution was so great that considering the amount of material available for analysis it is questionable whether there was any aluminum in solution at all. Considering, in addition, that the material examined was withdrawn from the upper intestine by a tube passed through the stomach throws even greater doubt upon the occurrence of aluminum in solution in the intestinal tract as a whole.

Recalling Tardieu's dictum (p. 43) "the concentration is of more importance than the dose," what is there to be said of the "irritating" action of such a concentration of the aluminum salts in the stomach? The individual who can differentiate in the mouth a solution of such concentration from any acknowledged bland solution or even from water itself is possessed of a hyperacuity of taste. In conjunction with the very great dilution of any aluminum that may pass into solution, it should further be appreciated that the gastrointestinal surfaces are of what may be termed a hardy type, that is to say, adapted to contact with food admixtures as ingested, without having undergone any physiological selection other than by taste in mastication to exclude a substance that might be "insidiously" irritating. It will be recalled that from beginning to end the gastrointestinal surfaces consist, not of cells that are adapted to exposure only to contents that are isotonic and of a certain fixed chemical composition, but on the contrary of cells that physiologically are adapted to contact with gastrointestinal contents of widely varying kinds that may be either hypertonic or hypotonic and that may present a wide range of chemical composition. The mucus secreted by the many goblet cells of the mucosa also affords protection by physically preventing too intimate contact between the mucosa and the contents within the lumen, by lubricating of hard material and thus avoiding mechanical irritation and also by chemical combination, if need be, with contents not adapted to intimate contact with cell protoplasm. This adaptation of the lining mucosa of the alimentary tract, for protection against the contents of the lumen, finds its counterpart in the construction of the epidermis of the body surface, which is thereby protective against various external agencies that acting directly upon the internal tissues of the body would be decidedly irritating. A familiar illustrative example that will suggest itself is the sensitiveness of the underlying tissue when not protected by the epidermis, such as occurs when a blister is formed and the separated external layer is removed. Precisely as the external skin affords protection against external agents, so does the mucosa afford protection against the varying contents of the lumen of the alimentary tract. Without the mucosa, as occurs in gastric and in intestinal ulcer, extreme sensitiveness is experienced.

It has only recently[1] been stated: "Aluminum compounds, being protoplasmic poisons, will poison the endothelia of the digestive tract with which they come in contact just exactly as I have seen aluminum compounds injure broken surfaces to which they have been applied." Further: "By the endothelia I mean the cells which line the digestive tract that come in immediate contact with the food and drink taken into the digestive tract." It is evident that the witness was here referring to what we have just described as the epithelia adjacent to the lumen of the digestive tract. That he was not ignorant of the protective function of the mucosa is apparent since, continuing, he stated: "These cells have not only the function of lining this tract, but they have a very important function in protecting the rest of the body" and, still further continuing, cites in illustration the protection from absorption of snake venom which through injured mucosa cells might gain entrance to the body and thus produce systemic toxic effects, a result from which the body is protected by the normal epithelia of the mucosa. No instance of the production of injury to the mucosa cells by the ingestion of baking powder food, either experimentally or otherwise, was given, though the context leaves no alternative but to conclude that the statements made were intended to convey the inference that such injury would or might be produced.

In answer to such a suggestion the writer must state without qualification that not only is there no evidence of any such injury ever having been produced, but that adequate knowledge of the subject is at hand to establish that there is no possibility that any such effect will or can be produced. One would be quite as justified to predicate pain and injury from the gentle touch of the body surface on the ground that the gentle touch on a raw surface is painful, as to conclude that a particular solution is harmful to the gastrointestinal mucosa because it is not a suitable solution in which to immerse isolated and unprotected cells. By the same reasoning the hydrochloric acid of the gastric contents would be recognized as harmful and its presence in the stomach in the normal amount a curse instead of a blessing.

It will be recalled that A. E. Taylor, in the Referee Board's investigations, reported (p. 117), with s.a.s. administered as such in large dosage, a tendency to produce "a somewhat characteristic type of gastrointestinal irritation with colic." Both he and Dr. Alfred Stengel, the physician in charge of the clinical observations, emphasized that the doses termed large were in reality very large. Indeed, not only were they at least 6 times the amount that ordinarily would be ingested, liberally partaking of baking powder food, and 2 or 3 times the amount by partaking very liberally, but also the s.a.s., being in a water soluble condition and being administered en masse, instead of being eaten in food throughout the day, would have its effects enhanced thereby at least several times in addition. Further, consideration of the fact that administration was made without the s.a.s. being decomposed and cooked in contact with food, would still further separate the experimental conditions from those that exist in actual consumption of baking powder food.

It will again be recalled that in a second series of experiments aluminum hydroxide was administered up to 3 times the maximum amount of aluminum in the s.a.s. series. The results of all the experiments led Dr. Stengel to attribute the colic and diarrhea to the alkali sulphates and led Dr. Taylor to conclude that in the absence of the sulphates colic and abdominal distress of less definite type only followed the ingestion of aluminum hydroxide in very large amounts.

That it was the conclusion of these investigators that such results were due to an abuse of the quantitative relation, as the writer commonly expresses it, and were not to be interpreted as an indication of the action of baking powder residue as it actually existed and was ingested in food, was clearly indicated in the conclusion of the Referee Board as a whole, to which Dr. Taylor subscribed, that the quality of foods made with an s.a.s. baking powder was not injuriously affected by reason of the presence of aluminum compounds therein. Likewise, Professor Long found the daily ingestion of the residues from 3 gm. of s.a.s. and soda combination objectionable from the production of cathartic action but concluded, since the practical daily consumption would be much below this amount, emphasizing that this was a practical and legitimate limitation, that it did not injuriously affect the quality of food. The Board also recognized that the residues from phosphate and from cream of tartar powders similarly produce a cathartic effect.

On the other hand, Professor Chittenden, working with amounts not only not so far removed from what might be ingested in food but also actually intimately mixed and cooked in food as would occur in practice, found no evidence of injurious effects whatsoever from s.a.s. baking powder foods.

Since quantitative limitations are necessary in considering the effect of any food or portion of food ingested, such limitations being determined by the amounts of actual consumption in practice, it is difficult to appreciate how there can be any proper exception to the conclusion of the distinguished investigators of the Referee Board because of their recognition of the differentiation of the effects of massive doses and the small amounts consumed in daily practice.

The only departure from such a rational viewpoint is to adopt an idealistic standard not applicable to food in general, which would condemn any substance in food that might produce ill effects if ingested in any amount by any person irrespective of age, activity or state of health. The writer cannot accept any standard that is not qualified by reason.

Inhibition of Enzyme Activity. Beginning with Ruttan's work in 1887 (p. 50), observations on the influence of aluminum compounds on enzymic activity in vitro were made by Mallet (p. 51), Pitkin (p. 49) Hehner (p. 52) and Bigelow and Hamilton (p. 55) during the period up ro 1900 when more importance was attached to such tesults than in later years. It was found in general that there was some inhibition of enzymic activity. There are several considerations that limit the value of such observations in their applicability to digestive processes in the body. One consideration is that the rate of enzymic action in vitro is affected by variations of conditions that are admittedly entirely innocent, so far as reflecting harmfulness to digestion in vivo is concerned. Thus, the presence of other food substances may markedly affect enzymic activity. Another consideration is the physical condition of the material under investigation, as, for example, whether crumb or crust be employed. Again, the particular proportions of enzyme and food is a determining factor; so much so that one proportion may show little or no effect on enzymic activity and another proportion a very definite effect. In recent years it has been quite generally recognized that relatively slight differences in enzymic activity in digestive rates in vitro do not reflect digestibility in vivo. The results of Long's digestive experiments (p. 111) led him to conclude that the residues in baking powder biscuits did not exert any action which seriously interfered with digestion and that in any event possible interference did not seem to be greater with biscuits made with s.a.s. powders than with other types of powders used for similar purposes.

The practical question presented by the influence on enzymic activities is whether there is inhibition of enzymic activity so as to interfere with the physiological utilization in vivo of the food product. The writer made carefully conducted investigations of this question (p. 67) by determining the coefficient of availability of a known diet in which bread made with an s.a.s. baking powder formed a conspicuously large proportion (65 per cent), comparing results with the coefficient of availability of a similar diet in which a control bread constituted the same proportion of the diet. It was found that the availabilities of the two diets were identical, indicating that there was no inhibition of secretory, motor or enzymic activities in the digestive tract so as to interfere to any degree with the utilization of the food in a physiological manner. Further, comparative availability determinations were made in a similar way of 2 subjects at the beginning and end of a nearly five-month period during which they had partaken daily of unusually large quantities of s.a.s. baking powder food (p. 85) and it was found that there had been no loss of capacity to utilize the diet in a physiological manner by reason of this prolonged period of ingestion of large quantities of s.a.s. food.

During the investigations of the Referee Board, attention was given to the influence of aluminum compounds on utilization of food in alimentation. It is evident that where catharsis was induced by the large quantities of alkali sulphate, comparative studies of food utilization could not be made, yet Prof. Chittenden's studies (p. 97), in which catharsis was not a prominent feature, directed particularly to fat and protein utilization, showed no interference when the baking powder residue was present in the food.

Likewise, studies of the stools by the Schmidt-Steele method, a procedure specially adapted to reveal undigested or partially digested food products, such as ordinarily occur where there is inhibition of the digestion processes, were made in the Referee Board investigations conducted by Professor Chittenden. His subjects were on diets for one hundred and thirty days which contained bread raised with s.a.s. baking powder, always in considerable and at times in very considerable amounts; yet when they were put on the special Schmidt-Steele diet required for the examination, the stools did not reveal the conditions during the aluminum periods that would show interference with the proper utilization of the food.

Considering all the evidence, the conclusion relative to digestive and absorptive processes in alimentation is that there is no interference by reason of inhibition of enzymic or other digestive activities, and that s.a.s. baking powder food is utilized by the body as is any other similar whole-some food.

The effect, if any, of food prepared with s.a.s. baking powder upon the bacterial flora of the intestines was given particular attention in the investigations of the Referee Board conducted by Prof. R. H. Chittenden (p. 97). There were of course daily variations in the data obtained, as those who have done much work upon the fecal bacterial flora know are regular and normal occurrences, but such variations had no relationship to the ingestion of the s.a.s. baking powder food and hence gave no support for the sug-gestion that such food tended unfavorably to modify bac-terial activities in the intestines. The observations by Prof. John H. Long, (p. 100), of the bacterial flora of the stools of subjects who partook of the assumed residues of s.a.s. baking powder, at times in very large amounts, although limited in scope did not lead this investigator to predicate any unfavorable effects of the ingested residues.

Both of these investigators studied the presence of prod-ucts of putrefaction in the stools, skatol, indol and phenol, also biliary pigment, and found no evidence of variation of the putrefactive processes in the intestines that had any relation to the residues ingested. Chittenden likewise found no significant variations of the urinary phenols and aro-matic oxy-acids. The earlier work of the writer in which urinary indican and ethereal sulphates were studied like-wise gave no evidence of the production of any gastrointes-tinal toxaemia.

The evidence of all the studies of the bacterial activities in the intestinal tract point, therefore, to one conclusion, namely, that there are no variations, even in individuals who have partaken of food prepared with s.a.s. baking pow-der or who have ingested the supposed baking powder residium in quantity over long periods of time, that would suggest or support any claims of the occurrence of abnormal bacterial growth or the production in abnormal quantity of the products of bacterial decomposition. Considering as well the normal utilization of the diet, the conclusion is clearly indicated that there is no production of what might be

termed intestinal indigestion, as a result of the ingestion of
baking powder residues in food even in the maximum
amounts in which such substances would be present in the
dietary. Any claims of the production of any such effects are
contrary to the evidence both from clinical and from scientific research.

3. THE CLAIM OF THE ABSORPTION OF ALUMINUM

Much emphasis has been placed upon the absorption of
aluminum into the blood from the alimentary tract.

E. E. Smith in his work on the hog (p. 70) failed to find
aluminum in the blood and organs of an animal who had
more than tripled in weight while on a diet consisting largely
of food leavened with a straight s.a.s. powder, used in double
the quantity indicated on the can. While there is probably
some justice in the criticism that the quantity of material
used in the analyses and the method of analysis employed
did not reveal amounts that would be found by the use of
larger quantities of samples and more highly specialized
methods of analysis as now employed, and particularly
more sensitive identification tests of the very slight residues
(0.1 mg. to 0.3 mg.), nevertheless, the results obtained did
clearly show that the claims then put forward, that there
was not only absorption but retention in the body in large
quantity as well, were not justified.

It is an adage that "qualitative tests precede quantitative
determinations." While this may not strictly apply to all
cases, yet it is an essential to all statements of quantity,
especially where the presence is equally at question as the
amount and of even greater importance, that the qualitative
identification of the substance be a part of the analytical
procedure. It is in the fact that there is no separation of any
aluminum that may be present and no opportunity for
adequate verification of its identity that the method advocated by Steel (p. 316) finds its most serious defect. This
objection is avoided in the Schmidt-Hoagland method,
developed by these investigators while on the research
staff of Dr. Taylor of the Referee Board. Since its publica-

tion, the method has practically supplanted all others in American research for the determination of aluminum in biological material. The first requirement for its proper employment is that reagents shall be used that are absolutely free from aluminum; not only to give value to the quantitative results, since a correction may be made for a very small weighable residue obtained in blanks, but chiefly to avoid the introduction of aluminum, which would vitiate the qualitative tests of the final residue. A second requirement for its proper employment is that the residue obtained be examined qualitatively and where necessary, quantitatively, for the presence of foreign admixtures. When it is considered that in the great majority of quantitative analyses of biological material, the amount of sample originally taken is necessarily large and in some instance may be as much as 500 gm. or more and that the amount of final residue obtained is exceedingly small, in many instances not more than a few tenths of a milligram, it will be appreciated how exceedingly difficult it is not to introduce a minute quantity of aluminum from the reagents and apparatus, and likewise how almost if not quite impossible it is to avoid the presence of some foreign material in the final few tenths of a milligram of residue actually weighed. The first difficulty, securing absolutely aluminum-free reagents, is correspondingly enhanced (not reduced as might at first seem probable) by the use of qualitative identification reagents of greater sensitiveness; the second difficulty, determining the character of the residue, qualitatively and quantitatively, makes the error proportionately greater, the smaller the amount of the final residue obtained. It is chiefly on account of these difficulties that some of the analysts who have given particular attention to the determination of aluminum in biological material have expressed grave doubts as to the justification of concluding that the few tenths of a milligram obtained indicate the amount or even the presence of aluminum in the original material (see Hilpert, p. 233; Howe, p. 230; Gephart, p. 234; Patten, p. 235; Myers, p. 195; Killian p. 195).

On *a priori* grounds there is reason to expect that aluminum will be present in the blood and tissues of higher animals, including man. Thus, the presence of aluminum in milk, both cow's and human, the value of traces of aluminum salts in animal diets composed of isolated food materials and the recent evidence from animal feeding experiments that aluminum salts are essential to reproduction can hardly be explained on other grounds than that aluminum is normally absorbed from the alimentary tract and that it exercises a true biogenic function in the chemical processes of the body. If this be so, then it only remains for the development of reliable tests of sufficient delicacy to demonstrate its normal presence and distribution. At the present time, there is some promise that the Atack method may constitute such a test, although its use, as indicated by available literature, has not yet been sufficiently extended to justify a conclusion in the matter.

A new line of investigation is now under consideration for the more accurate quantitative determinations of the amounts of aluminum in tissues. This is by use of the spectrograph, which produces a spectrum from ignition in an electric arc under standardized conditions. The method also serves for qualitative identification. The spectrum of aluminum is composed of lines each of which always has the same position. These lines are of different intensity. As the amount of aluminum volatilized in the arc is diminished the weaker of these lines disappear. As the amount is further decreased other and stronger lines also disappear. By determining at what concentration each different line disappears, it will be possible to determine the concentration of an unknown by the aluminum lines still remaining in its spectrum.

The presence of quantities of other elements in the unknown tends to blur the lines and form a more or less continuous spectrum. To avoid this it will be necessary to make the best separation of the aluminum possible, as has been done in the past, and apply the spectrographic method to the resulting ash. By determining thus the per cent of the ash that is aluminum, we will eliminate one of the great errors in present results. The spectrograph applied in a like manner to the reagents used will further insure the accuracy of the results. The ash in many cases is so small that a special technique must be developed to make the proposed method a success.

On the basis of present indications, notably the finding of aluminum in the liver of a "normal" dog by Balls (p. 225), using the Schmidt-Hoagland method, and in the tissues of two "normal" dogs by Gray, (p. 229), using the Atack method, considerable support is given to the view that aluminum is a normal constituent of the human* and animal body; and, further, in view of what has just preceded, that it likewise exercises a true biogenic function as does iodine, for example, whose importance to and even presence in the body until comparatively recently was unknown and its importance in the diet wholly unappreciated.

If it is found that aluminum is a normal and necessary constituent of the body, as is not at all impossible and of which there is from the evidence already cited some degree of probability, it will follow that it is normally absorbed, to some extent at least, from the digestive tract and that there is present in the body some method of procedure by which it is distributed to the parts where its presence is essential for the particular functioning with which it is concerned. As to the particular tissues which are the cites of its activities, the amount found (Gray, p. 229) in the thyroid and adrenals raises the question whether the selection for these tissues is merely accidental or has some as yet undemonstrated significance. Considering, too, Daniels and Hutton's findings relative to its value in reproduction, the question is presented whether it exercises some biogenic function by its presence in the testes or ovaries.

The question of its distribution at once calls to mind that in its chemical behavior aluminum deports itself in many respects precisely as does iron. It is because of this that its separation from iron in chemical analysis is possible by only a few chemical procedures. Iron, as we know, is absorbed in very small amount even though it be relatively abundant in the food and notwithstanding its wide distribution in the body. In its unnatural relation to isolated and unprotected cells in amounts greater than is required for biogenic func-

* Drs. Söldner and Camerer report[3] 0.3 gm. Al_2O_3 per kilogram body weight in infants.

tions it is quite as harmful as aluminum. Accordingly, is it not then possible that the same method by which iron is absorbed and distributed in the body, whereby the cells and tissues are protected from the direct effects of its ions, is also the method by which traces of aluminum are absorbed and distributed? This possibility is by no means wholly speculative. Balls (p. 227) has made certain observations from which, assuming their correctness as to the absorption of aluminum, he has concluded that there is a parallelism in the distribution of iron and aluminum, that aluminum, accompanies iron in its cycle (circulation in the body): bone marrow, corpuscles, liver, bone marrow, and that the aluminum exists, at least in the liver, in combination with proteins, exactly as does iron. So far as there is any evidence, then, the body employs the same method of absorbing and distributing aluminum as is employed for absorbing and distributing iron.

A vast literature has been created by the researches and speculations regarding the absorption of iron from the digestive tract. The present status of that controversy is summarized by Sollmann,[2] as follows:

"The increase of the iron content of the body in growing animals, and the equilibrium in adults, are sufficient evidence that food-iron is normally absorbed, metabolized and excreted. The controversy as to whether inorganic iron compounds can also be metabolized may now be considered settled in the affirmative: their fate is identical.

"All compounds of iron undergo some preliminary and probably unimportant changes during digestion. They are then *absorbed*, mainly from the duodenum, but also in decreasing order from the succeeding portions of the intestine. A part generally escapes absorption and is eliminated by the feces. The part which is absorbed passes into the intestinal epithelium in a dissolved condition, the organic compounds being at the same time changed into looser combinations. The absorbed iron may then pass directly into the capillaries; or through the stroma into the lymph, reaching the blood via the thoracic duct. The relative rôle of the blood and lymph absorption is not yet clear, but it seems that normally the main absorption is directly into the blood.

"From the blood, the iron is deposited as granules in an easily decomposed organic form (ferratin) in the cells of the hematopoietic organs, in the liver and red marrow, and particularly in the spleen, and to a less extent in the kidneys. In this form it remains until it is used or excreted. The *spleen* seems to be important in maintaining the iron reserve and in preventing excessive excretion.

"The *excretion* occurs mainly by the large intestines with traces in the urine, bile and gastric juice. The excreted iron is in firm organic combination.

"The *utilization* of the iron, its transformation into hemoglobin, occurs only as needed; so that the total quantity of hemoglobin does not rise above normal. The same is probably true of its utilization by other cells. An *excessive absorption* of iron results merely in an increase of the reserve stock."

It is difficult to state just how far the parallelism in the metabolism of iron and aluminum extends. Looked at from a quantitative viewpoint, it is undoubtedly true that the amount of iron required for the normal supply of ferruginous compounds of the blood and tissue is in excess, probably very considerably in excess, of the amount of aluminum necessary for normal growth and reproduction. Further than this, it is idle to theorize. Just as there is a normal absorption, storage and utilization of iron, so it may well be that there is a normal absorption, storage and utilization of aluminum, by at least some of the tissues. With the quantitatively lesser need for aluminum on the part of the body, it is not unlikely that aluminum is ordinarily taken up in correspondingly lesser amounts from the alimentary tract than is iron; and, further, that it is stored in the body in lesser amount, an amount in excess of some certain quantity being eliminated either by the liver, through the bile, or in the intestinal secretions, not an unusual route of elimination of certain metallic elements, either normal to the body such as calcium, or foreign to the body such as the heavy metals. That the body does not store aluminum in any appreciable quantity seems clearly established by the finding of, E. E. Smith (p. 90) and of A. E. Taylor (pp. 119, 124, 126) that it is eliminated, as near as can be determined, in the same amount in which it is ingested in the food.

Aluminum in the Blood or Tissues. Considering the results of analyses of the blood and organs of the body of the higher animals or of man by the methods available, observations of the presence or alleged presence of aluminum in the blood or tissues fall into various classes:

(1). Those in which there was no prior feeding of aluminum compounds, excepting as they were present as natural ingredients of the food.

(2). Those in which there was prior administration of soluble compounds of aluminum.

(3). Those in which there was prior administration of insoluble compounds of aluminum.

(4). Those in which there was prior feeding of food prepared with s.a.s. baking powder either (a) where the amount of baking powder used in the food was excessive; or (b) where the indicated amount of baking powder was used and the amount of such food fed was excessive.

(5). Those in which there was prior ingestion of the usual amount of food prepared with the indicated amount of s.a.s. baking powder.

Group (1). Analyses directed particularly to the detection of aluminum by available methods, excepting the Atack method, in the blood and various organs of higher animal and of man, not previously intentionally fed aluminum compounds, have in general failed to reveal its presence. This is true of the analyses of the organs of the domestic hog by E. E. Smith (p. 74), of several dogs by Balls (p. 225) and others, and of some specimens of human blood. Exceptions have been the instances in which Balls reported 1.1 mg., as AlPO$_4$, in 513 gm. blood of a dog, and the detection of aluminum in milk and in the bodies of new born infants.[3]

It may be concluded, therefore, so far as the limited analyses indicate, that aluminum ordinarily is not present in blood and in the various organs of higher animals and of man in sufficient quantity to be detected by the gravimetric methods of analysis, although Ball's finding in the blood of one dog, (0.48 parts Al per million) leaves it unjustified to conclude, if the Schmidt-Hoagland method be regarded as reliable, that it may not be. With the more sensitive Atack colorimetric method, however, the present indications are (Gray, p. 229) that aluminum, independent of special feeding, is or at least may be present, widely distributed in the organs of the dog (1 to 21 parts Al per million).

Group (2). Aluminum of the blood and organs of higher animals and man previously fed soluble aluminum compounds.

For the purpose of definiteness, in considering aluminum administration it is desirable to have some standard by which comparisons may be made. Taking into consideration the Referee Board's estimate that the ingestion of aluminum from s.a.s. baking powder food would ordinarily amount to from 25. to 75. mg. Al per day, it would seem that a fair unit for the ordinary maximum ingestion would be 1 mg. Al per kg. body weight. As a matter of convenience, this will be referred to in what follows as the Referee Board or R.B. unit of aluminum ingestion. Obviously, where the combination powders are used in food, instead of the straight powders upon which the above estimate is based, the ordinary maximum amount ingested may be considerably less than the R.B. unit.

Steel (p. 221), by his adaptation of the A.O.A.C. method, following the administration of alum to dogs in single dosage corresponding to 100 to 400 mg. Al_2O_3, amounts, considering the weights of the animals, corresponding to 2 to 12 R.B. units, reported in the blood, taken from three to seven hours thereafter, amounts of aluminum ranging from 0.85 to 7.3 parts per million. When alum containing 100 mg. Al_2O_3 was fed daily for eight days to each of 2 dogs, the blood of one, weight 13.4 kg., receiving 3.9 R.B. units Al, was reported to contain 1.5 parts Al per million; of another, weight 10 kg. receiving 5.3 R.B. units Al, was reported to contain 2.3 parts Al per million.

Group (3). Analyses of the blood and tissues of higher animals and of man subsequent to the administration of insoluble compounds of aluminum.

In this connection, Mott's claims (p. 47) do not merit serious consideration. Ball's feeding experiments of 2 dogs with aluminum phosphate (p. 226) fall into this class. The Schmidt-Hoagland method of analysis was used. The bile and gallbladder, kidneys, bone and the bone marrow each yielded figures greater than 5 parts Al per million. The amount fed, 10 gm. daily for a period of a month to a 14 kg. animal, was of course a very large dosage, being 160 R.B. units, that is, 160 times the maximum amount ordinarily consumed in baking powder food.

Likewise, Leary and Sheib's observations on the administration of alumina to dogs and rabbits, in which absorbed aluminum, amount not determined, was reported, are illustrative. Here, the dosage was even greater, being from 100 to 3000 R.B. units (p. 248).

Taylor's observations (p. 126) that the blood of men, fed large amounts of aluminum compounds, did not contain aluminum probably also come into this class.

Group (4). Analyses of the blood and tissues subsequent to the feeding of food prepared with an s.a.s. baking powder, either (a) where the amount of baking powder used in the food was excessive, or (b) were the indicated amount of baking powder was used and the amount of such food fed was excessive.

The analyses of the organs of a hog reared on large quantities of s.a.s. baking powder food (p. 74) did not yield, by the method of analysis employed (p. 73), detectable amounts of aluminum.

Steel, feeding respectively 16.3 and 19.1 R.B. units Al, reported the presence of aluminum in the blood (p. 221) of 2 dogs in the amount respectively 2.1 and 1.4 parts per million.

Kahn, feeding 3 dogs from 12 to 28 per cent more than the R.B. unit over periods of fifty-two to sixty-one days, reported (p. 222) the presence of aluminum in the blood in amounts from 5.8 to 9 parts per million; also present in the bile, long bones, kidneys, liver, muscle and spleen in amounts ranging from 0.5 to 111 parts per million. In the urine of a dog fed 38 per cent more than the R.B. unit over a period of nearly three months, Kahn reported 9.5 parts per million Al in the urine. The A.O.A.C.-Steel method was used and the amounts of samples actually used for the determinations were so small that there is a reasonable question as to the value of the results.

Balls (p. 227) fed s.a.s. baking powder food to each of 2 dogs for twelve weeks in amount corresponding to about 10.6 R.B. units daily and reported results, by the Schmidt-Hoagland method, according to the following parts Al per million for the respective animals: Blood 0.3–1.6, urine, in bladder, 1.9–8.7, bile 5.0–66.5, liver 0.5–0.7.

Killian (p. 228), feeding a dog for forty-two days amounts of food made with a straight s.a.s. baking powder that led to the daily ingestion of 190 mg. Al, being 23 R.P. units, reported figures, by the Schmidt-Hoagland method, corresponding to an increase of aluminum in the blood serum from 0.4 to 1.4 parts per million, no increase in the blood-clot, and amounts in the organs corresponding in parts Al per million as follows: Gallbladder and bile 8.9, liver 0.85, kidneys 2.3, bone 3.3. He did not regard the amounts as dependable or even as indicative of the absorption of aluminum in any quantity, owing to the limitation of the method.

Pine and Howe (p. 230) examined the blood of men whose diets for sixty-two, fifty-four and thirty days, respectively, had contained (assuming the body weight to be 70 kg.), as s.a.s. baking powder food residue, 5.5 R.B. units Al. From large quantities of blood the Schmidt-Hoagland method of analysis yielded residues weighing respectively 0.9, 0.6, and 0.8 mg. as $AlPO_4$, corresponding to 0.3, 0.26 and 0.6 parts Al per million. Qualitative tests, however, led the analyst (Howe) to conclude that aluminum was absent from the first and third of these residues but was present in the second residue, possibly to the amount of 0.0003 mg. This would correspond to 0.0006 parts Al per million or 6 parts per ten billion.

Balls and Gitlow (p. 231) fed 3 men s.a.s. baking powder food containing (assuming the body weight to be 70 kg.) 5.9 R.B. units Al for twenty-three, forty-six, and sixty days respectively; and from the analysis of the blood, using relatively small quantities, obtained by the Schmidt-Hoagland method residues corresponding respectively to 0.27, 1.4 and 6.4 parts Al per million. The presence of aluminum was verified by qualitative tests. From the urine of the second and third subject, they similarly obtained residues corresponding respectively to 0.08 and 0.6 parts Al per million.

Hawk and C. A. Smith (p. 232) for six weeks fed 2 men (F and S) s.a.s. baking powder food containing (assuming the body weight to be 70 kg.), 4.9 R.B. units Al; and from the analysis of the blood, using moderately large quantities, obtained by the Schmidt-Hoagland method, residues corresponding in duplicate respectively to (F) 3.1 and 0.9, average 2.0; and (S) 0.6 and 1.3, average 1.0 parts Al per million. They did not regard the method as accurate for blood but from qualitative tests concluded that aluminum was present in small amount. From the urine it was similarly concluded that aluminum was present to the extent of 0.2 and 0.07 mg. per day.

Group (5). Analyses of the blood subsequent to the ingestion of the usual amount of food prepared with an s.a.s. baking powder used in the amount indicated on the container.

Hilpert (p. 233) analyzed by the Schmidt-Hoagland method blood obtained from 4 subjects who had habitually partaken of s.a.s. phosphate baking powder food in the customary daily diet. It may be assumed that they consumed not more than 1 R.B. unit of aluminum per day. On the basis of the Schmidt-Hoagland method, the amounts of residues obtained corresponded to 0.46, 0.37, 0.69 and 0.66 parts Al per million, respectively. Aluminum was indicated by qualitative tests, but might have come from the reagents.

Gephart (p. 234) likewise examined blood obtained in a similar way. His results corresponded to 2.8, 0.8, 2.4, 4.3, 4.2, and 0.0 parts per million, respectively. The qualitative tests were not stronger than those obtained on the blanks of the reagents and he could not conclude that aluminum was originally present in the blood.

Discussion. It is seen from the above results that there is no unanimity of opinion regarding the presence of aluminum in the blood and tissues following its ingestion. By far the largest amounts were reported by Kahn (p. 222), yet this investigator used the smallest quantities of samples for analysis and employed the least certain of the recent methods.

Again, even by the Schmidt-Hoagland method, Killian (p. 227) obtained residues that lost a very considerable proportion on solution and reprecipitation, showing that what was actually weighed was not wholly or even largely aluminum phosphate, since the latter only diminished 3 per cent in weight when subjected to the same operation of purification, a conclusion confirmed by Howe's observation (p. 230) that similar residues obtained by him were not aluminum phosphate. The situation seems to necessitate the conclusion, in which Hilpert, Gephart and Patten as well were in accord, that neither the A.O.A.C.-Steel method nor the Schmidt-Hoagland method yields results bearing upon the absorption of aluminum that are reliable. This is no reflection upon these methods as analytical procedures where appreciable amounts of aluminum are present, since they were not designed for what may be termed a micro-chemical procedure.

It seems safe to conclude that there is no considerable amount of absorption or accumulation of aluminum in the body but that the question of the absorption so as to be present in the blood to the amount of a few parts per million is at present undetermined. On *a priori* grounds, as previously stated, it may well be that there is slight absorption, especially when the amount ingested is relatively large. It remains for a more highly specialized method to demonstrate conclusively whether or not this is so.

4. THE CLAIM OF ALUMINUM AS A PROTOPLASM POISON

It has been stated that aluminum in combination is a protoplasmic poison. By this is meant that it is inimical to the well-being of unprotected living cells when brought into intimate contact with them. The suggestion carries with it the opprobrium that as a consequence of this it is a harmful substance in the food of man. What are the facts in the case?

It is generally recognized that unprotected living cells react readily to their environment. When this is favorable, they grow rapidly and reproduce; when unfavorable they readily succumb. Because they are unprotected from external agents, they rank low in the scale of biological values. They may succumb because of an unfavorable composition of the liquid medium in which they are maintained, be this of a wrong concentration, be there an absence of substances to sustain the cells, be there an accumulation of their waste products, or be there present substances that are unfavorable to their well-being. Under any circumstances, unless they are so constructed that they exist in a dormant state, they are usually short lived,—a matter of a few hours or days when they either divide into new cells or die.

In the higher animals and in man, living cells lead a different existence. They are a part of a more highly organized being, which incorporates in its structure not only differentiation and coordination of cell activities but, what is of great biological importance, protective measures whereby the existence of its individual cells is prolonged, as serves the best interest of the structure as a whole.

When, therefore, we learn that a given substance in too great concentration is inimical to the maintenance of individual, unprotected living cells, if we are concerned with its influence on the well-being of higher animals and man, we are confronted with the consideration of the protective measures possessed by such higher forms of life; and whether these are or are not adequate to prevent injurious effects. If they are not fully adequate, then it will follow that the given substance will be deleterious; if fully adequate, then, it will follow that it will be innocuous.

Almost everything is in a sense a protoplasm poison, that is, under some conditions may act unfavorably on unprotected cell life. This is true both of foreign substances and of the essential constituents of protoplasm itself. Perhaps to an extent it is true in a general way that substances that enter in relatively large amounts into the composition of protoplasm are less and those that enter in relatively small amounts into its composition are more toxic to protoplasm. Both iron and aluminum, assuming the correctness of the view that aluminum exercises a biogenic function in cell life, are among the constituents that ordinarily are present in relatively small amounts.

Chapters VIII and IX contain records of attempts to determine the influence of aluminum compounds upon the growth and well-being of isolated cells. It is unfortunate that most investigators have omitted comparative studies of the influence of compounds of iron. The data available seem to indicate that the inorganic salts of iron and of aluminum have much the same effects upon protoplasm and that aluminum may be declared a protoplasm poison only in the same sense and to the same degree as may iron. With either of these elements, the question is presented in what way and to what extent the body of higher animals and of man affords protection to its individual cells from any unfavorable effects which the salts of iron and of aluminum are capable of producing on unprotected protoplasm.

The claim has been put forward that since in solution aluminum reacts with protein to form insoluble aluminum protein combinations, it would coagulate protoplasm and thus impair its functioning in a sense different and to a degree greater than would non-coagulants. So far as applies to the action of aluminum compounds upon unprotected protoplasm there would seem to be some force in this view. Since salts of iron have the same coagulating effect, the same reasoning would of course equally apply to their action as protoplasm poisons. With both of these elements, however, the very reaction which enhances the toxicity to unprotected protoplasm exposed to solutions of their inorganic salts may act as a protective measure when these elements are ingested in higher animals and man. The first contact of the ingested food on entering the body is with secretions containing proteins, so that the soluble iron or aluminum, as the case may be, and the insoluble combination, if to some extent it gradually dissolves in the gastric juice, is at once brought into intimate contact with protein with which it may enter into chemical union, thereby becoming fixed. At no place, then, are body cells exposed to contact with inorganic salts of iron or aluminum but with the element which has already reacted with protein. The protective nature of this combination is well illustrated by the writer's findings with the agar cup test (p. 175), in which penetration of agar by salts of aluminum was prevented when the aluminum was inactivated by admixture with blood. A reaction, then, that has been regarded as enhancing toxic action upon unprotected protoplasm, is to be regarded as a protective measure in the relation of ingested aluminum to the cells of higher animals and of man.

It will be seen elsewhere that the conclusion is reached that, as they occur in food and when administered through natural channels, adequate protection is provided and that these substances exert no deleterious influence.

5. THE CLAIM THAT ABSORBED ALUMINUM HAS A DELETERIOUS EFFECT

The claim that absorbed aluminum has a deleterious effect upon the blood-corpuscles and produces injury, notably to the kidneys and nervous tissues, with a tendency to the production of anemia and a lowered resistance to infections and disease, is based upon the results obtained by the subcutaneous and intravenous administration of salts of aluminum to higher animals. As a consequence of taking salts of aluminum into the body through natural channels, that is as constituents of food in any amount or for any length of time, no such injurious results have ever been observed, either upon higher animals or upon man. Obviously, the results of subcutaneous or intravenous administration to animals amply justify the conclusion that administered in this way, aluminum salts are harmful. Moreover such experimental observations are advantageous in a study of the action of ingested aluminum, by raising the question whether such effects ever occur as a consequence of the ingestion of aluminum compounds in food.

The fact that such injurious effects, following such ingestion, have not been found to occur, despite careful investigation, at once raises the question: What is the mechanism or mechanisms that intervene in oral ingestion whereby the injuries following subcutaneous or intravenous administration are not produced? It cannot be hoped fully to answer this question, but it will be helpful to recall some of the more obvious conditions, the so-called protective factors of the body, that intervene to prevent ingested ionizable salts of iron and of aluminum from producing the injury to blood and tissues that might result from direct contact or subcutaneous or intravenous administration.

Protective Factors. It hardly seems necessary to emphasize that protective factors are not to be credited to the ingenuity of scientists. The credit for their existence belongs to the power that created the structure and mechanism of living things. Wonderful it is to a degree that beggars description. For at least once, familiarity does not breed contempt. In the development of simpler into higher forms of life the protective factors have ever stood as essential features of the body's structures that have enabled such structures to survive. They have come into existence along with differentiation of structure and function as an essential part of such specialization without which the improved structure would not have survived and organic development would not have been possible. It is the protective factors that have made development, and now make our very existence, possible. They must certainly be taken into account for a proper consideration of the subject at hand.

A first factor of safety inherent in aluminum is the relatively limited solubility of its compounds. As with common table salt, any possibility of corrosive action is associated with the formation of relatively strong solutions. Under the conditions that exist with the ingestion of aluminum in food, either as a natural or an added ingredient, and particularly as the relatively insoluble baking powder residue, at the most only very weak solutions may be formed in the body. Upon this fact alone corrosive action in the body is excluded.

A second factor of safety inherent in the nature of soluble aluminum compounds is their property of reacting with protein to form relatively insoluble combinations. To some extent at least and probably to a considerable extent, this reaction occurs with the protein of food in its preparation, so that with the ingestion of alumed pickles or s.a.s. baking powder, more or less of the aluminum as well as iron present is already so combined. It is not held that this is a necessary condition, but it is mentioned because there are definite evidences that it is a condition that actually exists.

A third protective factor, and this is dependent on the body itself, is the mucous secretions of the alimentary tract. In the mouth, the saliva and throughout the tract from the stomach on, the mucus secreted by the mucosa not only prevents intimate contact of contents with living cells, but provides protein, more particularly phospho-protein, to combine with any dissolved, ionized iron or aluminum. While it is to be expected that concentrated solutions would overpower this provision for protection, there is every reason to believe that any solution resulting from either element in food would be so very weak that the protective mucus would be fully adequate to overcome it.

A fourth protective factor is the change of reaction when the gastric chyme passes into the duodenum, whereby compounds of iron and aluminum that may be dissolved are thrown out of solution. While it is admittedly true that the intestinal mucosa may absorb undissolved particles in very limited amount, it is generally held that, while solution tends to promote absorption, precipitation from solution acts in the reverse direction, tending to diminish and even practically prevent it. To what extent protein in solution exercises a solvent action on aluminum protein salts, or whether it does so at all, we do not know; nor has it ever been shown that the changes the chyme undergoes result in the formation of colloidal suspensions. If one is seeking support for the claim of any considerable amount of absorption of aluminum from the intestinal tract, emphasis will be given to the possibilities of such conditions. Until there is exact direct information available to either support or disprove such an occurrence, its existence or absence may be determined on other evidence. However, it is clear that on the whole the tendency in the intestines is to establish conditions unfavorable to absorption.

These factors, however, are not sufficient to disprove the absorption of aluminum, since much the same conditions exist in regard to iron, and evidence establishes the fact that iron is absorbed, particularly from the duodenum. However, iron is absorbed through the selective action of the epithelial cells, rather than between the cells, the avenue of absorption of metals in general.

While, from analogy with iron absorption, there is even a considerable degree of probability of the exercise of such a protective factor, it cannot be said on the basis of direct observation to have been established.

A fifth protective factor is the rate of absorption. Under any conditions, if absorbed, aluminum would pass through the intestinal walls into the blood relatively very slowly. Iron is not absorbed from the stomach. It is fair to infer that aluminum, if absorbed, would follow the channel of absorption of iron. As it would ordinarily take from two to four hours for the stomach to pass on into the duodenum the entire gastric chyme from a particular meal, any absorption of aluminum would consume a period of hours. Hence, even without considering the time it was retained in the epithelia lining the lumen, it would enter the blood very, very slowly, at a rate certainly hundreds and probably thousands of times less rapidly than when administered intravenously in experimental studies. This is true even on the assumption, which is contrary to fact, that aluminum ingested in food would be entirely absorbed. It requires no argument to appreciate that, assuming absorption, the rate would be exceedingly slow and that this might protect the body from untoward effects that would follow its rapid entrance into the blood.

Still another protective factor, the sixth in our list, is the highly albuminous character of the blood plasma and of any serous fluids in which the cells within the body are immersed. The agar cup test (p. 174), to which reference has already been made (p. 332), illustrated crudely the principle that ionized salts of aluminum react with proteins becoming thereby less active. In this condition it is fair to believe they are no longer chemically or physiologically active, as are the inorganic salts. It is probably true that they enter into protein combination before they reach the blood. However, it would seem that the fact that the medium in which they are contained is richly protein would tend to maintain this relatively inactive combination. It may do more than this. It is not at all unlikely that the aluminum-protein combination, ordinarily insoluble, is soluble in the plasma because of the plasma proteins, being thus carried in the portal circulation to the liver.

The liver may be regarded as another, the seventh protective factor. It is held that iron is stored in the liver in a firmer form of protein combination, the ferratin of Schmiederberg, if you wish. Balls has concluded that aluminum is similarly stored. His claim that it replaces the iron of ferratin and thereby becomes injurious cannot be taken seriously, since he has shown neither that because of such presence of aluminum there is diminution of iron nor that there is not an abundance of the protein moiety of ferratin to combine with both aluminum and iron. It would seem that this is a particular form of combination of either or both aluminum and iron for their storage in the liver in which form they are not inimical to protoplasmic activities and from which they may be converted into other combinations, as the needs of the body require.

That the needs of the body for iron are greater than the needs for aluminum seems indicated, not alone by the wide distribution of iron to practically all cells, but by the fact that iron storage in the liver is ordinarily greater than that of aluminum, the latter being present in such small amounts as to have generally escaped detection. Hence, if the future supports the present indications that aluminum exercises a biogenic function in the body, then it may be expected that with the development of more exact and delicate methods of analysis, it will be demonstrated that in some amount aluminum is a normal constituent of liver tissue. However, it is generally conceded that aluminum is not stored in the body in any considerable amount but if absorbed is rapidly eliminated, largely in the bile; possibly as well through the walls of the lower bowels. This rapid elimination, the fact that it is not cumulative, may be regarded as an eighth protective factor in the relation of aluminum to the body.

Having now sketched briefly some of the protective factors which intervene between the ingestion of aluminum compounds and action upon the cells and tissues of the body, it will be appreciated that there are reasons why oral ingestion is not followed by the inimical effects that follow intravenous administration. It becomes a question of whether these intervening protective factors are or are not adequate. It is generally conceded that for iron ingestion they are adequate; that is, that the ingestion of iron in food does not produce the inimical effects that would follow its intravenous administration and that therefore it is harmless—a conclusion to which, of course, one is forced by the fact that iron is essential to the body and must be ingested as a constituent of food. With aluminum, it has been amply demonstrated that aluminum compounds in food do not produce the effects that follow its intravenous administration, from which it follows that as with iron there are adequate intervening protective factors. Yet by a form of reasoning which it is hard to appreciate, it is held by some that, nevertheless, because such compounds produce injury when administered through unnatural channels, they are necessarily injurious in food. To satisfy their reasoning, which to others seems so clearly in error, they predicate that the injury is "insidious." That claim will be considered in the following section 6.

In connection with the subject of this section, some attention should be given to the results of investigations by Seibert (p. 244). She demonstrated marked blood degenerative changes from intravenous administration over a period of time of an amount of soda alum equivalent to approximately 1 or 2 R.B. units (exact weights of rabbits were not given). To other rabbits, there was administered by mouth in capsules, soda alum equivalent to 2½ or 5 R.B. units, and as a result of this latter, the degenerative changes observed from intravenous administration were not obtained (p. 244). However, during the period of oral administration there was a lowering of the hemoglobin to the extent of a few points, always, however, within the normal variation of hemoglobin in rabbits. This was regarded as "merely suggestive" of beginning blood degenerative changes, similar to the effects of intravenous administration. Obviously, the justified interpretation is that under the experimental conditions the oral administration did not demonstrate the degenerative changes of intravenous administration. The effects of oral administration have been amply and independently investigated on man by three members of the Referee Board (Chittenden, p. 94; Long. p. 99; Taylor, p. 114), the results clearly demonstrating that the effects of intravenous administration are not produced by oral ingestion, some of the reasons for which have been set forth briefly in the beginning of this section.

Claims of injurious effects upon the kidneys, nervous system and other organs have chiefly if not wholly been predicated upon the observations of the effects of subcutaneous and intravenous administrations of large doses of aluminum tartrate to frogs and rabbits (Siem p. 47; Dölkin). Not only are effects from oral ingestion not in any way indicated by the results of such intravenous administrations, but it is even not justified to conclude from the experiments to what extent, if at all, the aluminum ion is the cause of the injurious effects noted, since it is now known (p. 238) that the intravenous administration of tartrate causes marked necrosis of the renal cells through which the tartrates are eliminated. In any event, there is no indication that degenerative changes produced by administration through this unnatural channel are to any degree produced by aluminum compounds, as they are ingested in food. On the contrary, there is ample evidence that they are not so produced.

6. CLAIMS OF RETARDATION OF GROWTH AND DIMINISHED FERTILITY AND FECUNDITY

The claims that ingested aluminum compounds may produce acute toxic symptoms, retard growth and diminish fecundity with abnormality of the offspring have been based upon observations on animals fed large quantities of food leavened with excessive amounts of s.a.s. baking powder.

E. E. Smith (p. 70) fed 2 hogs for eighty-two days from the time of weaning, during which they received s.a.s. baking powder food, so as to ingest on the average 50 R.B. units of aluminum. During this time they gained in weight, respectively, 242.6 and 262.8 per cent. No ill effect of any kind was

produced. Two control pigs similarly fed, only without
s.a.s. as the leavening agent (sodium bicarbonate and hydrochloric acid were used), gained in weight respectively, 236.6
and 319.9 per cent. This was interpreted as normal growth
for the baking powder pigs. One of the control pigs was a
little lighter in weight at the outset, which accounted for
his slightly greater increase in weight, as compared with the
other three. He was not a runt pig, as he has been characterized by critics.

Two groups of 3 rats each were fed *ad libitum*, the one
s.a.s. baking powder bread, the other control bread, each
as above. The baking powder group gained in weight 487 per
cent in ninety-two days, the control group 539 per cent.
The difference in weight of the 2 groups seemed to be due to
one unusually large control animal. The experiment was
interrupted by the death of a control rat. The baking powder
animals displayed no evidence of ill health. The observations
were obviously too limited to justify any conclusions.

The observations of Leary and Sheib (p. 248), in which
puppies and rats were fed enormously excessive quantities
of aluminum hydroxide (equivalent to 100 to 3000 R.B.
units) have, as the authors themselves declare, "no special
bearing on the question of the behavior of aluminum compounds as used in the diet of man."

Dr. Gies and coworkers (pp. 249–271) fed rats an exclusive
diet of food made with excessive amounts of s.a.s. baking powder and interpreted the results as indicative of acute aluminum toxicity, diminished growth, diminished fecundity and
abnormality of the offspring. The aluminum ingested was
on the average at least 20 R.B. units, probably more, that
is, at least twenty times the maximum amount ordinarily
ingested by man from baking powder food. A certain relatively small number of the animals showed what the investigators called aluminum toxicity. Hawk and C. A. Smith (pp.
271–274) and E. E. Smith (pp. 277–284) failed to obtain any
"acute toxic effects." As was clearly pointed out by Steenbach (p. 287) and Mendel (p. 289), the symptoms observed
by Gies and coworkers were those of nutritional deficiency
and did not prove any aluminum toxicity whatsoever. More-

over, in an analytical consideration of the diet of Gies and coworkers, the author showed (p. 274) its marked deficiencies. Hence, not only was the production of aluminum toxicity contrary to the observations of Hawk and Smith and of E. E. Smith, but the interpretations of the observations by Gies and coworkers on the basis of their own results were unjustified.

Further, the diminished fertility, fecundity and abnormality of offspring observed by Gies and coworkers and by Hawk and Smith were fully explained by the nutritional deficiencies of the diets. (See Steenbach, p. 287; and Mendel, p. 289.) The animals fed excessive amounts of s.a.s. baking powder residues in food grew more slowly in the experiments of Gies and coworkers, Hawk and Smith and E. E. Smith. That they ate less than the animals fed the other foods was proved by the data of these experiments. An explanation of this has been offered by E. E. Smith (p. 286), in the unpalatability of excessive amounts of s.a.s. baking powder residue. Gies attempted to explain this on the ground that the animals were less healthy and hence ate less, citing in support of this view that they did not eat relatively more of the cheese and cabbage fed on Sundays. As all the animals were fed deficient diets, it is probable that in each instance they ate their full capacity on Sundays and that the s.a.s. group did not eat more because of the limitation of their capacity. It is clearly shown, then, that there was no justification, in the animal experiments cited, for a conclusion that the s.a.s. baking powder residues in any way interfered with growth, fecundity or well-being of offspring, excepting that in the enormous amounts present, at least 20 times the maximum amount of ordinary ingestion by man, such residues rendered the food distasteful. Hence they grew less because they ate less.

However, even if some degree of unwholesomeness had been indicated, which is contrary to the fact, the conclusion that baking powder food ingested by man might be unwholesome would be unjustified owing to the enormously greater quantity fed to the animals during the experiments. From any point of view unwholesomeness was not indicated by the experimental results obtained.

CONCLUSION AS TO CLAIMS

It will be noted that in all the evidence set forth to substantiate the various claims that so-called alum baking powders are injurious to health, not a single instance has been cited of the production of injury to animal or man from the ingestion of baking powder food as consumed by man. Considering the efforts made to establish the claim that such food is injurious or at least unwholesome, the failure to bring forward instances of ill effects resulting from the ingestion of such food is equivalent to an admission that such instances cannot be proved. Early in the trade propaganda to convince the consuming public that serious results followed the consumption of such food, "blind" newspaper accounts of specific instances of injury appeared in the public press. When such accounts were traced to their source, it was found in every case that such claims of injurious effects were without foundation.

The failure to substantiate the claims of actual injury has with the passing of time altered the claims of the opponents till at the present time "injurious" has been qualified to "unwholesome" and "demonstrable" to "insidious" effects. The right to claim insidious effects exists only when the fact of such insidious effects are established. Insidious ill effects may not be claimed merely because of inability to demonstrate actual ill effects. Nor, when the natural protective resources of the body have been overpowered in experimental study, either by the introduction of the substance of investigation through unnatural channels or by the quantitative abuse of administration through natural channels, may it be claimed that the effects produced represent to any degree the results of ingestion in the absence of such abuses. In such instances mal-effects result because of the overpowering of the protective resources of the body. It does not at all follow that any mal-effects are produced when the effects are well within such protective resources. The exercise of the protective functions is normal and physiological and may not without reason be characterized as abnormal or pathological and constituting an "insidious" injury. Such claims, based merely upon inability to demonstrate inimical effects, are tantamount to the admission that ill-effects are unproved.

SUMMATION OF EVIDENCE THAT ALUMINUM IS NOT INJURIOUS

The evidence in support of the conclusion that compounds of aluminum, as they are ingested in food from the employment of s.a.s. baking powders as leavening agents, do not render such food injurious to health, may be briefly summarized, in part, as follows:

1. Aluminum is a normal ingredient of food. This element is so widely distributed and is relatively so inert that its presence in food has commonly been disregarded. Nevertheless, it is so generally present and in such quantity that in a particular diet (pp. 14–21) it may be present in an amount several times greater than what would be introduced by the ingestion of food prepared with an s.a.s. baking powder (p. 37), there being 6 R.B. units in the particular diet cited.

2. Aluminum is not a foreign element in living matter, but is present as an essential constituent. This has been conclusively shown for plant life (pp. 176–181). In the preparation of a diet of isolated food substances, so constituted as to supply a balanced ration suitable for animal maintenance and to promote animal growth, salts of aluminum have been included in the supply of inorganic constituents. Moreover, the investigations of Daniels and Hutton (p. 293) give independent indication that aluminum is an essential constituent of food for the growth and reproduction of animals. This conclusion is in harmony with the observations that it is a normal constituent of cow's and human milk, which findings can only be explained by its presence in the animal and in the human body. Moreover, its presence in the newborn infant (p. 322) and its detection in animal tissues by the use of a particular method (p. 229), give direct indications that it is a normal and probably regularly occurring constituent of the animal body. It is strongly indicated, therefore, that aluminum is a normally occurring constituent of the animal and human body and that in the processes of maintenance, growth and reproduction it exercises a true and essential biogenic function.

The assigning of a biological rôle to aluminum adds interest to the observation of Stoklasa[4] that the colored forewings of beetles and all colored feathers are relatively rich in aluminum and his conclusion that the aluminum ion plays a definite rôle in the formation of the pigment.

3. Aluminum in the body deports itself much in the same way as does iron. In general iron and aluminum are similar in their chemical behavior. Under unnatural conditions, i.e., either when introduced into the body through intravenous or subcutaneous administration, or when ingested in massive doses, especially in soluble and ionizable combinations, they produce much the same effects. As a natural ingredient of food, it has not been established that either of these elements is harmful or objectionable. Indeed, as regards iron, it is recognized that it is an essential ingredient of food. Evidence is accumulating that aluminum is likewise an essential ingredient of food. It has been shown that iron is stored in the body in loose combination with protein, the so-called ferratin. Evidence has been presented that aluminum is similarly combined. In absorption, circulation and storage, combinations such as iron protein and ferratin protect the body from any untoward effects that would follow its unnatural introduction in soluble and ionizable combinations. There is every indication that aluminum, so far as it is absorbed, is similarly combined and that the body is similarly protected from any untoward effects which its soluble and ionizable combinations might produce. There seems to be this difference between iron and aluminum; that iron is required by the body in greater quantity and that it is accordingly stored in the body in greater amount. Aluminum is required in lesser quantity and is stored in relatively smaller amount, any quantity in excess of a certain very small amount being rapidly eliminated, chiefly by the liver through the bile, possibly by secretion into the bowels. Aluminum certainly does not accumulate in the body in any appreciable quantity.

Efforts have been made to liken the action of aluminum in the body to that of lead. It resembles lead in its action within the body only in the same way and to the same extent that the action of iron resembles that of lead, excepting that both iron and lead are cumulative and aluminum is not cumulative.

4. That extended scientific research has shown that food leavened with s.a.s. baking powders is not injurious when ingested by animal or man.

The earliest systematic investigations were those conducted by E. E. Smith (Chapter v). They showed that the ingested aluminum of such food did not accumulate in the body and that it was eliminated quantitatively in the feces; that such food was utilized by the body (digested and absorbed) in a normal manner and to a normal extent and that it did not give rise to disturbances of alimentation. Further, that even when ingested in relatively excessive amounts over experimental periods of nearly five months, there was no disturbance of general health, at the end of the research period the subjects presenting the same digestive utilization of the food and the same metabolic phenomena and the same clinical indications of general health and vigor, as at the outset of the observations.

Twelve years subsequent to the completion of the early investigations by the author, the Referee Board of Scientific Experts of the United States Government announced the conclusions of their investigations of this subject (p. 93) which fully substantiated the conclusions the author had reached and the opinions he had expressed. The Referee Board was an official branch of the Government, composed of the most eminent and qualified American scientists on the subject of nutrition, selected after consultation with the heads of America's leading institutions of learning, and appointed by the executive order of the President of the United States, Theodore Roosevelt, to investigate questions relating to the food supply of the American people and to advise the authorities, entrusted with the enforcement of the pure food laws of the government, of the Board's conclusions, for the guidance of these authorities in their interpretation and enforcement of the laws. It is safe to say that no body of scientific investigators in this field of knowledge was ever more competent or more impartial in its conclusions than was the Referee Board. Further, no similar problem was ever studied more thoroughly and intelligently than in the investigations conducted independently by the three members of this board to whom this particular subject was assigned. The investigations covered a period of several years. For this work each of the investigators had his own laboratories, properly equipped, and staffs, including medical supervisors, highly qualified scientists in the essential fields convered by the work, trained technicians and suitable experimental subjects. The report of the investigations, given in brief abstract, Chapter VI, occupied 2833 pages. In the abstract the extent and methods of investigation, as well as the results obtained, are sufficiently set forth to give a comprehensive appreciation of the basis of the Board's opinions and advice. As a matter of fact, the work covered practically every phase of alimentation, absorption, metabolism and elimination, as well as observations of the well-being of the subjects over long periods of time during the ingestion by man of s.a.s. baking powder residues and baking powder food. The entire Board concurred in the opinion rendered to the Secretary of Agriculture, relative to added aluminum in food, as given in full, (pp. 93–94), substantially to the effect that the use of s.a.s. baking powder, as a leavening agent in food did not affect injuriously the nutritive value of such food, did not reduce the quality or strength of such food and did not contribute any poisonous or other deleterious effect, so as to render such food injurious to health; adding only the qualification of the production of catharsis by the sodium sulphate and the production of occasional colic by the ingestion of large quantities. As a matter of fact, this last effect was not obtained from the ingestion of food but only from the administration of aluminum compounds in large excess of what would be ingested in baking powder food.

When Professor Chittenden, a distinguished member of the Board, was recently called before the Federal Trade Commission, he testified that no work reported by any scientist since the report of the Board had led him to modify or qualify the opinion which he expressed as a member of the Board and that he held the same opinion today that he had formerly written and to which he had subscribed as a member of the Board.

5. In spite of the extent of the scientific researches that have been conducted, covering both widely varied phases of the occurrence of aluminum compounds in food, and also the effects of aluminum under biological rather than practical conditions, the aspect of the subject that will perhaps appeal most strongly to the general reader and that supports as well the conclusion that such food is wholesome, is this one thing, namely the effect of s.a.s. baking powder food upon the millions of people who for nearly half a century have daily ingested such food, consuming at the present time, it is safe to say, 300 tons and probably more per day. The weight to be given the facts regarding such effects was ably set forth by the Court in State of Missouri vs. Whitney Layton (p. 158). Despite more than a quarter of a century of research by the opponents of s.a.s. baking powders, the facts remain today as they were then. On the basis of theoretical and general knowledge, certain experts maintain that serious results might follow the consumption of food leavened with s.a.s. baking powders. However, they have brought forth no "information or knowledge of any recorded instances in which functional disorders or disease or impairment of the digestion and general health had resulted to any human being," from the ingestion of food prepared with such powders. On this basis alone we may well state as our own the impartial opinion then expressed, made stronger by the passing of time:

"In the mind of the Court this fact, considering the enormous proportions to which the (so-called) alum baking powder industry has grown in this country, and the length of time such baking powders have been in use, stands as a stone wall against the deductions of the most eminent scientists who presented their theories on the part of the prosecution. I am unable to find in the evidence in this case any just ground for a ruling that (so-called) alum baking powders, of themselves, when used in the preparation of food are in any wise less wholesome than any other variety of baking powder."

BIBLIOGRAPHY

Bibliography for Chapter I

1. MELLOR, J. W. A Comprehensive Treatise on Inorganic and Theoretical Chemistry. N. Y., 1924, v, 154.
2. SMITH, E. E. Deleterious ingredients of food. *Science*, 1909, n. s. xxx, No. 773, 569–571.

Bibliography for Chapter II

1. LANGWORTHY, C. F. and AUSTEN, P. T. The Occurrence of Aluminum in Vegetable Products, Animal Products and Natural Waters. N. Y., 1904.
2. MYERS, C. N. and VOEGTLIN, C. Soluble aluminum compounds, their occurrence in certain vegetable products. *U. S. Weekly Public Health Rep.*, 1914, xxix, Pt. 1, Nos. 1–26, 1625.
3. Hearings, Docket 540, U. S. Federal Trade Commission vs. Royal Baking Powder Co.
4. PENNEY, M. D. Alum in flour and bread. *Chem. News*, 1879, xxxix, 80.
5. BLYTH, A. W. Foods: Composition and Analysis. London, 1888, 177.
6. TELLER, G. L. *Arkansas Station Bull.*, 1896, xlii, 75–77.
7. YOUNG, W. C. Alumina as a natural constituent of flour. *Analyst*, 1888, xiii, 5.
8. RICCIARDI, L. Sulla diffusione dell'allumina nei vegetali. *Gaz. Chim. Ital.*, 1889, xix, 150.
9. CHURCH, A. H. On the occurrence of aluminium in certain vascular cryptograms. *Proc. Roy. Soc. Lond.*, 1888, 44, 121.
10. JOHN, Cited by A. H. Church.
11. MAYER, A. Lehrbuch der Agricultur Chemie, Ed. 3, 1886, 280.
12. RITTHAUSEN, H. Neue Analysen der Ashen einiger Lycopodiumarten. *J. F. Prakt. Chem.*, 1853, I, lviii, 133.
13. AROSENIUS, Cited by A. H. Church.
14. KNOP, W. Kreislauf des Stoffes. 1868, 263.
15. MAIDEN, J. H. and SMITH, H. G. On a natural deposit of aluminium succinate in the timber of Grevillea robusta. *R. Brit. J.* and *Proc. Roy. Soc.*, New South Wales, 1895, xxix, 325.
16. SMITH, H. G. Aluminium the chief inorganic element in a porteaceous tree, and the occurrence of aluminium succinate in trees of this species. *J. & Proc. Roy. Soc. N. South Wales*, 1903, xxxvii, 107.
17. YOSHIDA, H. Chemistry of lacquer (urushi). *J. Chem. Soc. Lond., Trans.*, 1883, xliii, 472.

Bibliography for Chapter III

1. LÜBBERT and ROSCHER, *Pharm. Centralhalle*, xii, No. 39, 545.
2. LUNGE, G. and SCHMID, E. Über die Verwendbarkeit das Aluminiums zu Feldflaschen und anderen Gefässen. *Ztschr. f. ange. Chemie.*, 1892, 7.
3. RUPP, G. Ueber die Verwendbarkeit des Aluminiums zur Herstellung von Gebrauchsgegenständen für Nahrungs- und Genussmittell. *Dingler's Polytech. J.*, 1892, cclxxxiii, 19.
4. BALLAND, M. Sur l'aluminium. *Comp. rend.*, 1892, cxiv, 1536.
5. OHLMÜLLER, W. and HEISE, R. Unter Verwendbarkeit des Aluminiums zur Herstellung von Ess- Trink- und Kochgeschirren. *Arb. a. d. k. Gsndhtsamte*, 1893, viii, 377.
6. PLAGGE, DR. and LEBBIN, G. Ueber Feldflaschen und Kochgeschirre aus Aluminium. *Veraff. a. d. geb. d. Militär-Sanitätswesens*, 1893. No. 3, 1–100.
7. DONATH, E. Zum Verhalten und zur Anwendung des Aluminiums. *Dingler's Polytech. J.*, 1895, ccxcv, 18.
8. FRANCK, L. Gebrauch und Abnutzung von Aluminiumgeräthen im Haushalte. *Chem. Zeitung*, 1897, xxi, ii, 816.
9. MANSFELD, M. Aluminium-Geschirre. *Ztschr. f. Untersuch. d. Nahrungs- u. Genussmittel*, 1904, viii, 765.
10. VON FILLINGER, F. Über das Verhalten von metallischen Aluminium in Berührung mit Milch, Wein und einigen Saltzlösungen. *Ztschr. f. Untersuch d. Nahrungs- u. Genussmittel*, 1908, xvi, 232.
11. BARILLÉ, A. Attaque lente de l'aluminium par les eaux gazéifiées. *J. de pharm. et chim.*, 1912, ii, 110.
12. EDITORIAL. Some kitchen experiments with aluminium (from the Lancet Laboratory). *Lancet*, Jan. 4, 1913, 54.
13. EDITORIAL. Culinary and chemical experiments with aluminium cooking vessels. *Lancet*, March 22, 1913, 843.
14. BLOUGH, E. Personal communication.
15. Utz. Ueber die Verwendbarkeit von Aluminium in der Molkereipraxis. *Ztschr. f. ange Chemie.*, 1919, i, 345.
16. TRALLAT, Quoted by Dronilly, *Le Lait*, 1921, i, 228.
17. FRIEND, J. A. N. and VALLANCE, R. H. The influence of protective colloids on the corrosion of metals and on the velocity of chemical and physical change. *Trans. Chem. Soc.*, 1922, 468.
18. TINKLER, C. K. and MASTERS, H. The corrosion of aluminium cooking utensils. *Analyst*, 1924, xlix, 30.
19. GIES, W. J. Hearings, 1924, Docket, 540, U. S. Federal Trade Commission vs. Royal Baking Powder Co.
20. EDMUNDS, C. H. Hearings, 1924, Docket 540, U. S. Federal Trade Commisson vs. Royal Baking Powder Co.
21. CUSHNY, A. R. A Text-book of Pharmacology and Therapeutics, Phila., 1924, 670.
22. POULSSON, E. A Textbook of Pharmacology and Therapeutics (Dixon Translation), Lond., 1923, 479.

23. SOLLMANN, T. Manual of Pharmacology, Phila., Ed. 2, 1922.
24. RICHARDSON, W. D. The Current Significance of the Word Alum. Chicago, 1927.
25. JACOBS, B. R. Hearings, 1924, Docket 540, U. S. Federal Trade Commission vs. Royal Baking Powder Co.

Bibliography for Chapter IV

1. DEVERGIE, A. Médecine Légale, Paris, 1851, iii.
2. ORFILA, M. Rapport et experiences sur les effects de l'alun. *Ann d'Hyg.,* 1829, i, 235.
   Traité de médecine légale, Paris, 1848, iii, part 1, 190.
   Traité de toxicologie, Paris, 1852, i.
3. HASSALL, A. H. Food; Its Adulterations, and the Methods for their Detection, Lond., 1857, 281.
4. TAYLOR, A. S. Poisons in Relation to Medical Jurisprudence and Medicine. Phila., 1859, Am. Ed., 311.
5. STILLÉ, A. Therapeutics and Materia Medica, Phila., 1874.
6. RICQUET, DR. Un cas d'empoisonnement par l'alun. *J. de pharm. et de chim.,* 1873, 4 s., xviii, 333.
7. TARDIEU, A. Étude Medico-Légale et Clinique sur l'Empoisonnement, Par., Ed., 2, 1875.
8. PATRICK, G. E. Alum in baking powders. *Scient. Am. Sup.,* July, 1879, viii, 2940.
9. WEST-KNIGHTS, J. On the action of alum in bread-making. *Analyst,* 1880, v, 67.
10. MOTT, H. A. JR. The effects of alumina salts on the gastric juice in the process of digestion. *J. Am. Chem. Soc.,* 1880, ii, 13.
    Reprinted with additions, The Baking Powder Controversy, A. Cressy Morrison, 1907, ii, 1131.
11. SIEM, P. Ueber die Wirkung des Aluminiums und des Berylliums auf den thierischen Organismus. Diss., Reported by Kobert, R., *Schmidt's Jahrb.,* 1886, ccxi.
12. PITKIN, L. On the solubility of alumina residues from baking powders in gastric juice. *J. Am. Chem. Soc.,* 1887, ix, 27–30.
13. PITKIN, L. Report, 1889, The Baking Powder Controversy, A. Cressy Morrison, 1907, ii, 1375.
14. RUTTAN, R. F. The digestibility of certain varieties of bread: an experimental study of the alum question. *Trans. Roy. Soc. Can.,* 1887, v, sect. 3, 61.
15. MALLET, J. W. Experiments upon alum baking-powders and the effects upon digestion of the residues left therefrom in bread. *Chem. News,* 1888, lviii, 276.
16. HEHNER, O. On the influence of alumed baking powder on peptic digestion, with remarks on a recent prosecution. *Analyst,* 1892, xvii, 201–209.
17. BIGELOW, W. D. and HAMILTON, C. C. The influence of alum, aluminum hydroxide and aluminum phosphate on the digestibility of bread. *J. Am. Chem. Soc.,* 1894, xvi, 587.

18. Döllken. Ueber die Wirkung des Aluminiums mit besonderer Berück-sichtigung der durch das Aluminium verursachten Läsionen im Central-nervensystem. *Arch. f. exper. Path. u. Pharmakol.*, 1897, xl, 98.

Bibliography for Chapter V

1. Smith, E. E. Effects on digestion of food prepared by the use of an alum baking powder. *N. Y. Med. J.*, 1900, lxxii, 719.
2. The Baking Powder Controversy, A. Cressy Morrison, 1907, ii.
3. Unpublished data.

Bibliography for Chapter VI

1. Report of Referee Board of Consulting Scientific Experts on Influence of Aluminum Compounds on the Nutrition and Health of Man, 1914, U. S. Department of Ag iculture, Washington, D. C. (unpublished).
2. American Men of Science, Ed. 2, 1910.
3. Op. cit. Ed. 3., 1921.
4. Hearings, Docket 540, U. S. Federal Trade Commission vs. Royal Baking Powder Co.
5. Bulletin 103, U. S. Department of Agriculture (Professional Paper). Alum in Foods. 1914.

Bibliography for Chapter VII

1. Norfolk Baking Powder. *Analyst.* 1879, iv, 231.
2. West-Knights, J. On the action of alum in bread-making. *Analyst.* 1880, v, 67.
3. Hehner, O. On the influence of alumed baking powder on peptic digestion, with remarks on a recent prosecution. *Analyst*, 1892, xvii, 201–209.
4. Case of James James, on Appeal. *Analyst*, 1893, xviii, 152.
5. The Baking Powder Controversy, A. Cressy Morrison, 1904–7.
6. Opinion, in the Court of Quarter Sessions of Dauphin County, Penn., No. 61, January Session, 1910.
7. Hearings, Docket 540, U. S. Federal Trade Commission vs. Royal Baking Powder Co.

Bibliography for Chapter VIII

1. Smith, E. E. Deleterious Ingredients of Food. *Science*, 1909, n. s. xxx, No. 773, 569–571.
2. House, H. D. and Gies, W. J. The influence of aluminum-ions on lupin seedlings. *Proc. Am. Physiol. Soc.*, 1905, 19; *Am. J. Physiol.*, 1906, xv.
3. Hartwell, B. L. and Pember, F. R. Aluminum as a Factor Influencing the Effect of Acid Soils on Different Crops. The Presence of Aluminum as a Reason for the Difference in the Effect of So-called Acid Soil on Barley and Wheat. *J. Am. Soc. Agronomy*, 1918, x, No. 1.
4. Conner, S. D. Liming in Its Relation to Injurious Inorganic Compounds in the Soil. *J. Am. Soc. Agronomy*, 1921, xiii, No. 3.
5. Conner, S. D. and Sears, O. H. Aluminum Salts and Acids and Varying Hydrogen-ion Concentrations in Relation to Plant Growth in Water Cultures. *Soil Science*, 1922, xiii.

6. OSTERHOUT, W. J. V. (a) Injury, recovery and death in relation to conductivity and permeability. Phila., J. B. Lippincott Co., 1923.
(b) The Effect of Some Trivalent and Tetravalent Kations on Permeability. *Bot. Gaz.*, lix, 464–73.
7. Hearings, Docket 540, U. S. Federal Trade Commission vs. Royal Baking Powder Co.
8. STOKLASA, J. Uber den Einfluss des Aluminum-ions auf die Keimung des Samens und die Entwicklung der Pflanzen. *Biochem. Ztschr.*, 1918, xci, 137–223.

Bibliography for Chapter IX

1. HEILBRUNN, L. V. The Electrical Charges of Living Cells. *Science,* 1925, lxi, 236.
2. Hearings, Docket 540, U. S. Federal Trade Commission vs. Royal Baking Powder Co.
3. McGUIGAN, H. The pharmacology of iron and aluminum in relation to therapeutic uses. *J. Lab. & Clin. M.*, xii, No. 8, 1927.

Bibliography for Chapter X

1. BALLS, A. K. Dissertation, Columbia University. The occurrence of aluminum, and its absorption from food, in dogs. N. Y., 1917.
2. Hearings, Docket 540, U. S. Federal Trade Commission vs. Royal Baking Powder Co.
3. MALLET, J. W., Personal statement to W. J. GIES, August 27, 1909. See M. Steel. *Am. J. Physiol.* 1911, xxviii, 94.

Bibliography for Chapter XI

1. Hearings, Docket 540, U. S. Federal Trade Commission vs. Royal Baking Powder Co.
2. The Baking Powder Controversy, A. Cressy Morrison, 1907, ii, 1199.
3. LOEVENHART, A. S. Hearings, Docket 540, U. S. Federal Trade Commission vs. Royal Baking Powder Co.
4. SMITH, E. E. Deleterious ingredients of food. *Science*, n. s. 1909, xxx, 569.

Bibliography for Chapter XII

1. LEHNER, V. Hearings, Docket 540, U. S. Federal Trade Commission vs. Royal Baking Powder Co.
2. The Baking Powder Controversy, A. Cressy Morrison, 1907, ii, 1114.
3. STEEL, M. On the absorption of aluminum from aluminized food. *Am. J. Physiol.*, 1911, xxviii, 94–102.
4. MIXER, C. T. and DU BOIS, H. W. The Zimmermann-Reinhardt method for the determination of iron in iron ores. *J. Am. Chem. Soc.*, 1895, xvii, 405.
5. KAHN, M. On the absorption and distribution of alumized food. *Biochem. Bull.*, 1911, i, 235–244.
6. SCHMIDT, C. L. A. and HOAGLAND, D. R. The determination of aluminum in feces. *J. Biol. Chem.*, 1912, xi, 387–391.

7. Howe, P. E. A comparison of the method proposed by the Association of Official Agricultural Chemists as modified by Steel, with that described by Schmidt and Hoagland, for the determination of aluminum in organic material *Biochem. Bull.*, 1916, v, 158–164.

8. Curtman, L. J. and Gross, P. A study of the methods for the quantitative determination of aluminum in blood. *Biochem. Bull.*, 1916, v, 165–172.

9. Steel, M. The determination of aluminum in the presence of iron and organic matter. *Biochem. Bull.*, 1916, v, 173–182.

10. Smith, C. A. and Hawk, P. B. The determination of aluminum in biological material; a comparison of the methods of Steel (modified by Kahn) with the methods of Schmidt and Hoagland. *Biochem. Bull.* 1916, v, 183–188.

11. Balls, A. K. A direct test of the degree of accuracy of the Schmidt-Hoagland Method for the determination of aluminum. *Biochem. Bull.*, 1916, v, 195–202.

12. Atack, F. W. A new reagent for the detection and colorimetric estimation of aluminum. *J. Soc. Chem. Indust.*, 1915, xxxiv, 936.

13. Wolff, L. K., Voorstram, N. J. M. and Schoenmaker, P. Bepaling van kleine hoeveelheden aluminium. *Chem. Weekb.*, 1923, xx, 193.

14. Keilholz. Diss., Leiden. Cited by Atack.[12]

15. Hearings, Docket 540, U. S. Federal Trade Commission vs. Royal Baking Powder Co.

16. Balls, A. K. The occurrence of aluminum, and its absorption from food, in dogs. Diss., 1917, Columbia University, N. Y.

17. Gonnermann, M. Beitrage zur Kenntnis der Biochemie der Kieselsaure und Tonerde. *Biochem. Ztschr.*, 1918, lxxxviii, 401.
Zur Kenntnis der Biologie der Kieselsaure, Tonerde und des Eisens. *Ztschr. f. physiol. Chem.*, 1920, cxi, 32.

Bibliography for Chapter XIII

1. Orfila, M. Treatise on Legal Medicine. Par., 1848, Ed. 4, iii, Part 1.

2. Pereira, J. The Elements of Materia Medica and Therapeutics, Phila., 1852, Am. Ed. 3, Ed. by Joseph Carson, I.

2a. Webster, R. W. Potassium and Tartrates. Chicago, 1927.

3. Baer, J. and Blum, L. Ueber die Einwirkung chemischer Substanzen auf die Zuckerausscheidung und die Acidose, III. *Arch. f. Exper. Path. u. Pharmakol.*, 1911, lxv, 1–34.

4. Underhill, F. P. The influence of sodium tartrate upon the elimination of certain urinary constituents during phlorhidzin diabetes. *J. Biol. Chem.*, 1912, xii, 115–126.

5. Underhill, F. P., Wells, H. G. and Goldschmidt, S. A note on the fate of tartrates in the body. *J. Exper. M.*, 1913, xviii, 317–346.

6. Idem, 322.

7. Simpson, G. E. Urinary secretion of tartrates, following administration to animals. *J. Pharmacol. and Exper. Therap.*, 1925, xxv, 459.

8. Steel, M. On the absorption of aluminum from aluminized food. *Am. J. Physiol.*, 1911, xxviii, 94–102.

9 Hearings, Docket 540, U. S. Federal Trade Commission vs. Royal Baking Powder Co.

Bibliography for Chapter XIV

1. LEARY, J. T. and SHEIB, S. H. The effect of the ingestion of aluminum upon the growth of the young. *J. Am. Chem. Soc.*, 1917, xxxix, 1066–1073.
2. Hearings, Docket 540, U. S. Federal Trade Commission vs. Royal Baking Powder Co.
3. DANIELS, A. L. and HUTTON, M. K. Mineral deficiencies of milk as shown by growth and fertility of white rats. *J. Biol. Chem.*, 1925, lxiii, 143–156.
4. RUHRÄH, J. The soy bean as an article of food for infants. *J. A. M. A.*, 1910, liv, 1664.
5. OSBORNE, T. B. and MENDEL, L. B. The relation of growth to the chemical constituents of the diet. *J. Biol. Chem.*, 1913, xv, 311–326.
6. ANDEREGG, L. T. Diet in relation to reproduction and rearing of young. *J. Biol. Chem.*, 1924, lix, 587–599.
7. MATTILL, H. A., CARMAN, J. S. and CLAYTON, M. M. Nutritive properties of milk; effectiveness of x substance in preventing sterility in rats on milk rations high in fat. *J. Biol. Chem.*, 1924, lxi, 729–740.
8. MITCHELL, H. S. and SCHMIDT, L. Relation of iron from various sources to nutritional anemia. *J. Biol. Chem.*, 1926, lxx, 471–486.

Bibliography for Chapter XV

1. WELLS, H. G. Hearings, Docket 540, Federal Trade Commission vs. Royal Baking Powder Co.
2. SOLLMANN, T. Manual of Pharmacology. 1917, 763.
3. SÖLDNER and CAMERER, DRS. Die Aschenbestandteile des neugeborenen Menchen und der Frauenmilch. *Ztschr. f. Biol.*, xliv, 1903, 60.
4. STOKLASA, J. Über die Verbreitung des Aluminiums in der Natur und seine Bedeutung beim Bau- und Betriebsstoffwechsel der Pflanzen. Jena, 1922, 122–129.

INDEX OF AUTHORS*

* See also Bibliography, pp. 349–355.

INDEX OF SUBJECTS

PAUL B. HOEBER, INC., 76 Fifth Avenue, New York